# About Island Press

Since 1984, the nonprofit organization Island Press has been stimulating, shaping, and communicating ideas that are essential for solving environmental problems worldwide. With more than 1,000 titles in print and some 30 new releases each year, we are the nation's leading publisher on environmental issues. We identify innovative thinkers and emerging trends in the environmental field. We work with world-renowned experts and authors to develop cross-disciplinary solutions to environmental challenges.

Island Press designs and executes educational campaigns, in conjunction with our authors, to communicate their critical messages in print, in person, and online using the latest technologies, innovative programs, and the media. Our goal is to reach targeted audiences—scientists, policy makers, environmental advocates, urban planners, the media, and concerned citizens—with information that can be used to create the framework for long-term ecological health and human well-being.

Island Press gratefully acknowledges major support from The Bobolink Foundation, Caldera Foundation, The Curtis and Edith Munson Foundation, The Forrest C. and Frances H. Lattner Foundation, The JPB Foundation, The Kresge Foundation, The Summit Charitable Foundation, Inc., and many other generous organizations and individuals.

The opinions expressed in this book are those of the author(s) and do not necessarily reflect the views of our supporters.

# THE CURIOUS LIFE OF KRILL

# The Curious Life of Krill

## A Conservation Story from the Bottom of the World

Stephen Nicol

Foreword by Marc Mangel

**ISLAND**PRESS

Washington | Covelo | London

Library of Congress Control Number: 2017961065

Manufactured in the United States of America
10 9 8 7 6 5 4 3 2 1

*Keywords:* Island Press, krill, crustacean, Antarctic, Southern Ocean,
CCAMLR, euphausiid, sea ice, swarm, schooling, krill fishery, A-frame,
Association of Responsible Krill Fishing Industries (ARK), Antarctic Treaty
System (ATS), Biomass, Biological Investigations of Marine Systems and Stocks
(BIOMASS), Convention on the Conservation of Antarctic Seals (CCAS),
cod end, Discovery Investigations, echosounder, fast ice, formalin, genome, gills,
International Geophysical Year (IGY), International Whaling Commission
(IWC), Marine Stewardship Council (MSC), metazoan, Marine Protected Area
(MPA), pack ice, pelagic, phytoplankton, polynya, Scientific Committee on
Antarctic Research (SCAR), thermocline, zooplankton

# Contents

# Foreword

You are holding a remarkable book in your hands. One reason is the particular organism it discusses, the Antarctic krill—*Euphausia superba*. It is named "superb" for good reasons, some of which I mention here and all of which are discussed in the book.

You will also learn the scientific history of Steve Nicol, one of the leading krill scientists of our time, and about the history of the Commission for the Conservation of Antarctic Marine Living Resources (CCAMLR), the first organization to adopt an explicitly ecosystem-based approach to fisheries management. Clearly this book is about a specific place (the Southern Ocean), but it is as much an adventure of ideas as it is bound to location. Here's a little bit about krill, Steve Nicol, and CCAMLR.

Krill are crustaceans related to shrimp and prawns, but unlike their relatives krill live in the water column rather than on the ocean bottom. Antarctic krill are likely the most abundant (by biomass) animal on the planet; they inhabit entire oceans and use the whole ecosystem in the course of their lives. You'll learn that Antarctic krill form some of the largest animal aggregations and have an amazing biology that includes long life and large size, as well as the ability to shrink when times are tough. They can eat up to twenty percent of their body weight per day, and they are strong swimmers that can go wherever they want to go in the ocean. And, of course, krill are a terrific food source, powering the biggest animals (blue whales) on Earth. Indeed, most other organisms in the Southern Ocean either eat krill or eat something that eats krill.

Although we have studied krill for a long time, mysteries about them still abound. For example, why do we find them living and even mating at depths of six hundred meters (650 yards) below the surface? Did we only need to think earlier to search at deeper depths?

Your guide for this adventure, Steve Nicol, is a stellar scientist. Steve completed a PhD working on krill, and in 1987, after a couple of years of postdoctoral work, joined the Australian Antarctic Division. He has been deeply involved in krill research ever since. He also holds a master's degree in creative writing from the University of Tasmania, and the quality of writing shows on every page of the book.

Thus Steve has worked on krill for forty years and over that time has become one of the world's leading krill scientists. I have had the good fortune to know him for nearly thirty years. We first met during a working group concerning krill in 1989 (Steve describes that working group in the book).

Steve has been a leader in learning about live krill at sea and in the laboratory, where his team figured out how to grow krill successfully and get them to reproduce and swarm in the laboratory. These advances have allowed us to learn much about krill that we simply could not have determined from observations at sea. For example, after recognizing that about a quarter of the iron in the top twenty meters (twenty-two yards) of the ocean was bound up in krill, he pioneered iron fertilization experiments that broke new ground in our understanding of krill ecology.

Steve's personal story will also show you how science is done in practice (not in the idealized schoolbook manner that most people imagine), that obtaining more data often makes a problem more complex than simpler, and how industry and science can interact in positive and reinforcing ways.

You will learn in this book that krill have been the target of the largest fishery in the Southern Ocean for the past forty years. Because of their enormous biomass, there is not much concern about harvesting krill unsustainably, but rather about local depletion of krill, especially around the breeding colonies of birds and seals. This concern has been highlighted because of new technologies for krill fishing that Steve describes.

Those concerns led to the establishment of the Convention for the Conservation of Antarctic Marine Living Resources (CCAMLR). The story of CCAMLR is the third thread in this book, and Steve is perfectly positioned to tell it since he has been on the Australian delegation to the CCAMLR Commission for thirty years.

The CCAMLR treaty is the first ecosystem-based approach to the management of fisheries, thinking not only about the target species but also about dependent predators and the consequences of fishing for them. You will learn about the "krill flux" and the "krill surplus," associated, respectively, with movement of krill and the consequences of the near extinction of the great whales. And you will learn about the various krill products and their associated hype. Indeed, there were more than eight hundred krill-related patents filed between 1976 and 2000 for food, health products, and pharmaceuticals, such as enzymes for cleaning wounds.

Steve tells the story of the development of precautionary measures (starting in 1991), which were first applied in October 2010. We have gone from a situation in which there were no limits anywhere on krill fishing in 1991 to one in 2009 in which any krill fishery anywhere in the Antarctic was subject to strict limitations. Although the future of krill and krill harvesting is uncertain, Steve is optimistic about both.

Early in the book, Steve writes that virtually all science is interpretation: scientists share facts, and their challenge is to convert data into knowledge, which depends on how we understand and interpret the data. As scientists, we have the most impact when we embed our facts and data in a compelling story. And Steve Nicol has a very good one.

Bon voyage.

Marc Mangel
University of California, Santa Cruz, and University of Bergen

# *Preface*

So, this is all we are, lunch! Goodbye krill world.
—Will and Bill the krill, in *Happy Feet 2*

I like krill—and there are only a handful of people on the planet who can genuinely make this claim. This is not because krill are not likable—quite the contrary—but because most people don't know enough about them to have an opinion. Bringing up krill in a conversation usually guarantees a blank look in return. After enduring years of vacant stares, my resolve to educate the world about krill grew stronger. I would write a book, *the* book, on krill. I set out to answer the question, why are krill still largely unknown or misunderstood?

The world is home to an entire universe of animal species, many of which have enthusiasts who champion their cause. All biologists, given half a chance, will work hard to convince unbelievers that *their* particular species of interest is a critical cog in the great machine of life. So why would my book on krill be different, and what unique message do I bring in the following pages? My aim is to convey the importance of this not-so-small crustacean in the Antarctic ecosystem as well as in the global scheme of things.

I hope, after reading this book, you will no longer exclaim, "I didn't realize they were so big!" when you see them on a TV screen or en-counter them at an aquarium or perhaps even in the wild. Maybe you will join me in my crusade to banish the description of krill as "tiny shrimp-like crustaceans" or "zooplankton." Who knows, you may even feel moved to praise their delicate and feathery beauty, and their tremulous and sensitive behavior.

If I have done my job well, then you should be capable of engaging in an educated discussion about the role of krill in marine ecosystems and of venturing an opinion about the sustainability of the krill fish-ery and whether it is well managed. Above all, I truly hope that you find yourself liking krill as much as I do.

I began working with krill nearly forty years ago—and I have been one of those lucky scientists who has been able to study my pet spe-cies throughout my career. I was convinced early on that krill were vital to marine food chains, but I quickly found that my fellow ma-rine scientists downplayed the essential animal nature of krill (and of most marine organisms) and thus misunderstood their complicated life cycles.

Over the years, I have also encountered an almost complete lack of awareness from the general public, journalists, politicians, and even my fellow scientists about the true nature of krill. For an obscure animal, this might be excusable, but krill, as we shall see, are crea-tures that challenge humans for the title of most abundant animal on Earth and are vital in maintaining the health of the oceans.

There is very little accessible information on Antarctic krill, and much of what is available is highly technical in nature. My aim in writing this book is to provide an account of krill that can be read and understood by a wide audience. This is not a scientific book, rather, it is an attempt to synthesize what we know about krill from a range of sources, some scientific, some more eclectic. The end-product is *my* interpretation of all the facts that I have available, molded into a (hopefully) readable story about a fascinating creature—Antarctic krill.

As it is my interpretation, it probably differs quite radically from a similar approach that might be taken by any of my colleagues. That's okay. Science operates by developing stories (we call them conceptual models) about the natural world. These stories are meant to be plausible accounts about how the system in question might work, but they are never intended to be the final word. Our stories provide the basis for the next generation of studies, which, in turn, test the feasibility of our ideas, resulting in a revised or entirely different story. This is the way that we make progress and how we increase our confidence in our ability to understand an organism or system. Disagreement is an essential part of this process, and if my fellow scientists disagree significantly with my story, then hopefully they will develop an alternative story that better synthesizes the available information, and the field of krill biology will have made significant progress.

This book provides a broad-brush approach to one of the biggest and certainly the best-known species of krill: Antarctic krill, scientifically (and appropriately) named *Euphausia superba*. Although this is the best-studied species of krill, there are still vast holes in our understanding of how they live and survive in a deep ocean that is ice-covered and dark for over half the year.

Their habitat provides a clue to our ignorance. It is difficult to carry out research on krill except in summer, and then only for short periods of time and under uncomfortable conditions. Studying other species of krill in more temperate waters is easier, and such studies can provide clues to the great success of the krill family in general. But because Antarctic krill have a unique prominence in the Southern Ocean, we need to study them there to understand how they thrive in an environment that is unbearably harsh for humans.

This book is also an examination of how we study and interact with animals in the open ocean and why there are still so many oceanic mysteries yet to be resolved. I will spend some time in this book explaining how we obtain knowledge about Antarctic krill. Hopefully this will explain why there are still so many gaps in our knowledge and why so much of the krill story is speculative despite the decades of study and volumes of scientific papers.

My long research career has given me a good grasp of the huge scientific literature that has featured krill, and I will use this knowledge to tell the story of Antarctic krill and their environment. We will never fully understand krill (or any other animal), but by presenting what we know in a coherent story we will be able to assess the overall state of knowledge and then plan, strategically, for the future directions of research.

But this book is not merely an ode to a special creature, it is an answer to the need for accessible information on a species that affects not just marine ecosystems at the bottom of the world but also our day-to-day human lives. Products such as krill oil, "sourced from the pristine waters of the Antarctic," are appearing on drugstore shelves with increasing regularity, and we consumers need to know more about their source, how they are manufactured, and how their fishery is managed, now and into the future.

My focus is on the conservation biology of krill, but this is not a book with a conservation agenda. Hopefully the neutral approach I have tried to adopt will enable conservationists, scientists, fishermen, and politicians to understand the issues. The oceans are changing in ways that will dramatically affect krill, and we will expect politicians and managers to make decisions based on credible scientific information. I hope this book will help.

*Chapter One*

# Oceans of Krill

*Euphausia*—true shining light (Greek)

It was night, and the trawl deck of the Canadian research vessel *Dawson* was an oasis of light in the inky summer darkness. The ship rolled in the gentle swell, revealing a phosphorescent wake as the stern rose and fell. From the A-frame a single wire stretched taut into the floodlit water, which parted around it in a V-shaped wake as the ship moved slowly forward. At the appropriate signal, the ship's hydraulics groaned into action, and the cable was slowly retrieved from the depths, dripping seawater from the pulley block. A cry of "Sight!" indicated that the net was close to the surface, a pale billowing ghost just visible in the subaquatic greenish light. The drizzling net emerged from the water and was hoisted overhead like an elongated mosquito net. As the wire reached the zenith of the A-frame, the net hung suspended above the glistening deck. The scientific crew sprang into action, using a seawater hose to wash the contents into the net's cod-end—the small container at the end of the net that holds the catch.

Satisfied that every inhabitant of the deep had been flushed into its final resting place in the cod-end, the lead researcher undid the clasp that fastened the cylindrical cod-end to the long, flimsy body of the net. I joined the crowd of eager students on deck as contents were decanted into a plastic tray. "Bugs!" dismissed the professor, as the students enthusiastically poked the rapidly dying catch with tweezers.

A fine-mesh net towed through the water has been standard equipment for sampling the smaller inhabitants of the ocean for more than two hundred years. Because the ocean depths are still one of the planet's great unknowns, there is always a sense of excitement when the contents of a net are heaved onto the deck for examination. The animal life collected in a net has an unforgettable odor—uniquely marine, slightly musty, yet not entirely unpleasant. The animals themselves can vary widely, from wriggling worms to delicate jellies, often crushed and punctured by their unanticipated journey to the surface, to robust amphipod crustaceans that ricochet off the walls of the specimen tray, tiny flealike copepods that twitch their way through the water, and sea snails that appear strangely out of place in the open ocean.

Some larger animals caught my eye—shrimplike creatures that had clearly endured an uncomfortable trip to the surface. These were krill. They lay on their sides, their swimming legs going through the motions, their bodies turning opaque as they slowly died under the glare of the ship's spotlights.

This was my first oceanographic cruise, and my first encounter with living-but-soon-to-be-dead krill. I watched as the tray was whisked away and its contents poured into a jar and topped with formalin—an evil-smelling chemical concoction used to preserve specimens for later analysis. Biological oceanography in the 1970s was essentially this process: dragging a Victorian-era device behind the ship, dispassionately sorting its contents on deck, preserving specimens of interest, and tossing the rest overboard. The exercise was repeated throughout the area of ocean being studied. We gave little thought to the diversity of the sea life we collected, or how the creatures lived and behaved. They were "bugs," zooplankton whose lives were ranked by size, weight, and relative abundance. Through the barrel of a microscope, one animal suspended in formalin looked much like the next.

On a calm day, later that same year, the sun glinted off the gently heaving waters of the Bay of Fundy on the Atlantic coast of North America. Distant flocks of birds hovered above the calmer waters of the oily slicks that meandered across the surface. Our small fishing boat slowly motored toward the signs of biological activity, and I scanned the water for further signs of aquatic life. In the calm, emerald water, rays of sunshine penetrated several meters into the depths of the bay, brightly illuminating surface features, but deeper objects remained indistinct and ghostly. The dense green light of the surface layer distorted the colors of deeper objects. Coastal debris—clumps of seaweed, leaves, twigs, plastic bottles, buoys, and other rubbish associated with fishing communities—had accumulated in the slicks. But below this film there was movement.

Several meters down I could make out a whitish cloud, which gradually turned pink, then red, as it rose to the surface. The cloud became a swarm of thousands of animals with elongated, tapered bodies, swimming in unison and headed in the same direction as if they had a common purpose. As the school neared the surface film, the water turned a rich ochre-red, and I could make out individual krill. Each was about three centimeters (roughly an inch) long, separated by one centimeter from its neighbors. Each probed the water ahead with a set of four antennae, backed up by a pair of bulging black eyes. Krill bodies rippled as their swimming legs propelled them smoothly through the water, their movements perfectly synchronized with those of others in the school. As the krill swarm reached the surface, the school extended their antennae out of the water, and the calm surface rippled from thousands of crustaceans sensing the air. As I watched, mesmerized, this mass of crustaceans became a living, brick-red raft, writhing on top of the water's surface. As if on cue, thousands of krill flipped their muscular tails and leaped clear of the water, falling back like a shower of pink raindrops.

This abundance of densely packaged protein did not go unobserved by the rest of the food chain. The gradual ascent of the krill to the surface was tracked by schools of squid that darted into the aggregations from below, grabbing individual krill with their tentacles, then shooting backward into the ocean's murky depths. Shoals of herring flashed through the krill schools, disrupting their orderly structure and leaving a snowfall of silver scales in their wake. Birds—shearwaters and gulls—squabbled over the smorgasbord of seafood, devouring predator and prey alike. Loud exhalations heralded the arrival of mammals: dolphins and giant fin whales, intent on a meal of krill, fish, and squid, lunged into the krill swarms with mouths agape.

As our boat drifted amongst the swarms of krill and their splashing predators, I watched one long, ribbon-like school of krill moving purposefully past the boat, antennae forward, black eyes prominent, and tails ablur as their swimming legs rhythmically and in unison pushed water away. Each krill maintained a set distance from its neighbors, and the school moved as one—most of the time. As the school passed the boat, one krill broke ranks and swam toward the watching humans, momentarily pausing a meter away. It appeared to check us out before turning around at high speed to rejoin the school.

This first encounter with living krill and their intriguing range of complex social behaviors left me captivated, and the spectacle became my second lesson in biological oceanography. Although routine specimen collection is still scientifically necessary, it became clear to me during this encounter that dead krill under the microscope can provide only part of the scientific story. Preserved specimens allow us to identify and classify anatomical features and conduct important measurements and observations. But just as learning about human life involves more than studying cadavers, learning about krill involves drifting alongside them, entering their world, learning their rhythms and observing their interactions with their swarm-mates and their predators. My attitude toward life in the oceans changed in an afternoon; no longer could I view krill as mere bugs; I was beginning to appreciate their complexity and their beauty.

Given the difficulties of studying life in the open ocean, most marine scientists learn about their subjects by second-hand methods: samples collected in nets, traces on an echosounder readout, or blurred images on an underwater video. There is little capacity to personally observe most animals in the open ocean, or to witness their interactions with the rest of the ecosystem. There are exceptions, however, and in my search for a PhD project, I stumbled across one of them in the Bay of Fundy, a dynamic body of water that separates Nova Scotia from New Brunswick and Maine. According to local ornithologists, krill rose to the surface here in vast numbers each summer and became the food of vast flocks of seabirds. Why? The conventional scientific wisdom of the day conjectured that the massive tidal range, up to eight meters in places, brought krill passively to the surface, where seabirds and whales lined up to feast on these highly visible, calorie-dense treats. But nobody had studied the krill swarms in detail, and I was determined to figure out the mystery. Why would millions of krill take such a suicidal journey in broad daylight?

I spent several idyllic summers searching for an answer, messing around in a small boat with an affable fisherman, chasing krill, catching fish, and watching whales—eventually answering that question to the satisfaction of my doctoral dissertation committee. The answer, like the answer to so many biological mysteries, was sex; the krill were at the surface to mate and lay their eggs. But I could only arrive at this conclusion after several years of sampling, observations, and long hours in the laboratory hunched over a microscope.

My guide and mentor was Ray Thurber, an archetypical wise old fisherman who knew the Bay of Fundy intimately in the old-fashioned way, without the aid of modern instrumentation. He could read patterns in surface waters that changed with the tides and could take me to any location of the bay with barely a glance at his depth sounder. GPS had not yet been invented, and if it had, he wouldn't have needed it. Ray taught me that a fisherman's livelihood depends on his ability to catch fish, and so, like scientists, fishermen devise, and test, elaborate theories about what structures life in the sea. If they're successful, they catch fish. If they're not, they must refine their theory or come up with a new one.

Ray and I set out in his small cape island fishing boat each morning, just after dawn, to search the bay for krill. We were accompanied by a flotilla of fishing boats, their skippers all convinced that early morning was the best time to catch fish. On our way out to sea we passed a different flotilla coming in from a night of fishing, their skippers equally convinced that night was the best time to catch fish. Fishermen, like scientists, interpret the results of their investigations in many ways and modify their sampling strategies accordingly.

Ray was always happy to put his theories to the test. I asked his advice on the currents that swirled around the ledges where the krill regularly appeared, and he was sure he knew where the currents came from and where they flowed to. We would conduct experiments, throwing drifters into the water to track the current flow. Sometimes they ended up exactly where Ray predicted, sometimes not. "There's another theory shot to hell!" he would mutter from behind his corncob pipe when his predictions failed to pass muster.

Fast-forward fifteen years, and I was at the negotiating table as a member of the Australian delegation to the international commission that regulates the fishery for Antarctic krill in the Southern Ocean, where I was proposing the establishment of limits on the catch of krill off a vast area of East Antarctica. The catch limits we had calculated were based on research I had initiated involving a lengthy seventy-two-day survey on the *Aurora Australis*, a ninety-five-meter (104-yard) Australian icebreaker. We had conducted the first-ever census of the krill population specifically designed to regulate the fishery. Our team of forty-two scientists and twenty-five crew used state-of-the art electronics, sampling gear, and advanced computing to estimate the size of the krill population. Although this was a dramatic change in scale from my previous sampling activities in Ray Thurber's small green boat, the motivation behind my multi-million-dollar research project was the same—to understand how krill live in their environment so that we could ensure their conservation. The scale and complexity of the operation had changed and so had the species of krill we were studying. This time it was Antarctic krill, the superb krill.

~

What are krill? Ask the average person on the street to describe a krill, and the most frequent response is a blank look. On rare occasions the response is "krill are the tiny shrimp-like crustaceans that whales feed on." Many imagine that krill are microscopic, like water fleas or phytoplankton. Few appreciate their real size, which is far from microscopic. If all the animal inhabitants of the ocean from the largest whale to the most minuscule invertebrate are lined up based on size or weight, krill fall in the middle of the pack. Next to their seafaring brethren, they are average in size—not large but not microscopic. Not even tiny.

Antarctic krill, one of eighty-five krill species, were first noticed by whalers in the Southern Ocean. The bellies of the giant whales they hunted were filled almost exclusively with what whalers variously described as shrimps, squillae, animalcules, and insects. Krill had identity issues from the start. Whalers had known of the existence of krill from their observations in the North Atlantic, where several species are abundant and were frequently sighted on the feeding grounds of the great whales.

The word *krill* is understood to mean "young fish" in Norwegian, but I was informed by a Norwegian colleague that the term is actually an onomatopoeia, a word formed to replicate the sound of millions of krill pattering on the water as they jumped clear of the surface—behavior I had witnessed firsthand in the Bay of Fundy. This surface swarming behavior, exhibited by many species of krill, was another indication of their existence. Since krill are generally only found deep in offshore waters, they are usually seen only by fishermen and whalers and others who venture far from land.

Krill go by many names in the languages of maritime nations. Gaelic fishermen knew them as *suil dhu*, which means "black eyes." In Japan they are known as *esada* or *okiami*, and fishermen in the English-speaking world often refer to them as red bait. But most people have no need of a word for animals they never see and rarely hear of.

In the language of science, however, krill belong to the taxonomic order of Euphausiids, and several of their eighty-five species are abundant in oceanic and some coastal areas around the world. Not surprisingly, Antarctic krill, first scientifically described in 1855, are found in the waters around the frozen continent. It is, without doubt, the most abundant, ecologically and economically important, and best-studied Euphausiid species, and consequently is the focus of this book. From here on, I will use the term *krill* to refer to Antarctic krill. If I am dealing with other species, such as my first scientific subject, North Atlantic krill (*Meganyctiphanes norvegica*), I will clearly indicate it.

Like *sheep*, or *fish*, the name is both singular and plural, so I will use *them* when referring to the species as a whole, and *it* when referring to an individual.

Krill are mostly transparent, when alive, with splashes of contrasting red and green, large spherical black eyes, and an elongated, streamlined shape that tapers to a pointed feathery tail. The green comes from the algae they eat and the red from special pigmentation spots that can expand and contract, making their shells a lighter or darker shade of red. When viewed en masse krill can turn the ocean blood-red.

All species of krill also have electric-blue light-emitting organs that dot the body, providing a spectacular light show, particularly when freshly collected. The overall effect is startling, and those who see living krill for the first time are often struck by their translucent beauty. When viewed closely, they are undoubtedly crustaceans, distant cousins to the prawns and shrimps that are so familiar in the displays in fish markets.

The Australian Antarctic Division in Tasmania established a research aquarium especially for the study of Antarctic krill. The aquarium quickly became a favorite of tour groups who grew to appreciate the beauty of free-swimming krill, including visitors as diverse as schoolchildren, politicians, and wildlife celebrities, such as David Attenborough and Jacques Cousteau. This was often the first time they had seen living krill. Inevitably, comments began with, "I didn't realize they were so big!" This has led to my lifetime quest: to quash the misconception that krill are microscopic organisms. Adult Antarctic krill can reach a length of over six centimeters (roughly two and a half inches) and can weigh up to two grams (0,07 ounces).

Over the years, frustrated by constant references to the microscopic size of krill, I made what many have considered an unwise decision. I had a krill tattooed on my left arm as a handy illustration of the appearance and actual size of krill that I could flourish when needed. That was the theory anyway. Unfortunately, my tattoo artist, though undoubtedly talented, was unable to render my favorite crustacean at the requisite size, and he took a few liberties with its anatomy too. As a result, when I roll up my left sleeve I display a

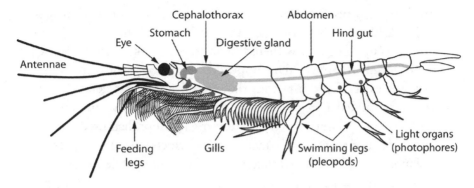

The major anatomical features of Antarctic krill. (Drawing by Marcia Rackstraw)

rather terrifying lobster-like creature about twice the size of a krill while declaring "krill look something like this. They are quite a bit smaller but are definitely not microscopic!" Not quite what I had in mind, but it's still a guaranteed conversation starter.

It would be overstating the case to suggest that the beauty of krill has been the subject of poetic attention, or that literature and works of art are littered with depictions of krill. That said, a few lyrical descriptions of krill do appear in unexpected places. One of my favorite early accounts of krill comes from a scientist who was working on the whaling stations in South Georgia in the 1930s:

> The "krill" is a creature of delicate and feathery beauty, reddish brown and glassily transparent. It swims with that curiously intent purposefulness peculiar to shrimps, all its feelers alert for a touch, tremulously sensitive, its protruding black eyes set forward like lamps. It moves forward slowly, deliberately, with its feathery limbs working in rhythm and, at a touch of its feelers, shoots backward with stupefying rapidity to begin its cautious forward progress once again. (F. D. Ommanney,1938)

Few modern writers have come close to capturing the essence of krill better than this.

In contrast, most modern descriptions are rehashed clichés and are rarely written from firsthand experience of living krill. This is not surprising because living krill are hard to find, and good images of live krill are scarce. An Internet search for krill more often than not turns up images of withered brown specimens practically reeking of formalin through the computer screen. Even worse, many depict species of other crustaceans that are mistakenly labeled as krill. I was once handed a brochure about the need to conserve and protect krill, but the sponsoring conservation organization had used a cover photograph of a swarm of squat lobsters, which have only a passing resemblance to krill. It is difficult to convince the public to conserve a species you can't correctly identify.

Krill are certainly crustaceans with a shrimplike appearance. They are differentiated from actual common shrimps and prawns in that they don't perch on the seafloor, but rather swim in the open ocean for their entire life—a pelagic, or "free-swimming," lifestyle. To swim constantly requires enormous inputs of energy as krill are heavier than seawater. This apparent drawback has not impeded krill from an evolutionary perspective, and they are among the planet's most successful inhabitants by any measure.

Krill also differ from other crustaceans in subtle ways, most of which could be spotted only by ardent crustacean scientists. For example, their gills are found outside their carapace (the largest part of the shell that covers their head and bodily organs). Sensory organs are strategically located on the front end of the body and consist of two pairs of antennae and large eyes. Krill also possess ten pairs of feathery feeding limbs, which they use to comb the water in a not-very-discriminating search for food particles, living and dead, animal and plant. This makes them highly versatile omnivores with a constant supply of food that allows them to swim, grow, and reproduce. Their tail is their powerhouse, almost pure muscle, which activates their five pairs of swimming legs to propel them forward, and it provides thrust for rapid backward escapes, often referred to as lobstering.

Krill rarely swim alone; they are most often found in swarms or schools containing astronomical numbers of companions, which, as anyone who comes from a large family can attest to, has both costs and benefits. Because they are average-sized, abundant, and aggregated, krill are a premium food source for organisms higher up the food chain. Entire marine ecosystems depend on krill to eat the truly tiny organisms, and in turn to provide themselves as concentrated protein for bigger animals. Krill are large enough to be seen from a distance by predators, and their swarms are large and dense enough to be easily detected by animals using echolocation, such as some species of whales. Krill swarms can be so extensive and visually striking that they can be seen from space.

It is difficult for krill scientists like me to avoid superlatives when describing Antarctic krill, because they are truly among the world's most astonishing animals. Krill possess distinctive features that truly set them apart from the humdrum life in the oceans. As an entrée to this book, I'll finish this introductory chapter by listing seven remarkable attributes of krill that will be elaborated on in later chapters.

*First, krill are possibly the most abundant animal on the planet.* It is a bold assertion that Antarctic krill could be the world's most abundant multicelled animal (metazoan), and I should clarify that (1) I refer to Antarctic krill as a single species, not a general group, such as ants or beetles; (2) I use biomass as the unit of measure, meaning the number of individuals of a species multiplied by their average weight; and (3) I make the (rather shaky) assumption that we can accurately quantify the biomass of any species.

With those caveats in mind, an analysis by the World Wildlife Fund in 2011 suggested that cattle have a biomass of 520 million metric tons, human biomass is around 350 million metric tons, and krill are a rather distant third, with 150 million metric tons. A more recent assessment of krill biomass has put the figure at around 400 million metric tons, so krill are still in the running for global biomass domination. Without a doubt, however, krill have the greatest biomass of any marine metazoan. Because the oceans occupy seventy percent of the area of our planet, this fact alone makes them globally important.

*Second, krill inhabit an entire ocean.* Estimates of the area of the Southern Ocean inhabited by krill range from nineteen to thirty-two million square kilometers (7,336,000 to 12,355,000 square miles), which is ten percent of the planet's ocean area, or roughly four times the land area of Australia. It is now apparent that Antarctic krill can also be found in all water depths—from the surface to the sea floor at 4,500 meters (4,921 yards) below. Living in this immense volume of water has extreme implications for krill, ranging from how they maintain connectivity between widely distant populations, to what physiological processes they require to cope with the daily pressure changes at different depths.

For scientists who study krill, the logistics are similarly complex, for it is impossible to access more than a small portion of their icebound habitat at any one time. Ascertaining population trends is the basis of ecology and conservation science, but understanding changes in krill populations, when they are found in such a vast and inhospitable area, is fraught with difficulty.

*Third, krill form some of the largest aggregations of animal life.* Krill live in three-dimensional swarms, or schools, that can stretch for twenty kilometers (approximately twelve and a half miles) and contain as many as three million metric tons of animals (roughly thirty trillion individuals). Not surprisingly, such swarms have been described as the largest aggregation of animal life on Earth by the arbiter of all big things—Guinness World Records. These vast dynamic clouds of krill are difficult to envision, though the film *Happy Feet 2* did a wonderful job of portraying what a krill swarm might look like, even if the constituent members of these swarms were slightly modified in the animation. The aggregating habit, the density of the swarms, and their sheer size make krill ecologically critical and economically important.

*Fourth, krill are much bigger than people think.* As I have already mentioned, krill, with a length of six centimeters (roughly two and a half inches), are far from being microscopic. Krill size has a number of consequences for their visibility as prey and how they are valued by humans. It is difficult to mount a krill conservation campaign, for example, when the world assumes they're no bigger than microscopic plankton. Conservation efforts have historically enjoyed more success for large animals, such as whales, elephants, and tuna; interest in saving species wanes for those on the smaller end of the spectrum.

Thus depicting a small animal as being smaller than it actually is compounds the problem.

*Fifth, krill have a unique biology.* So many aspects of the biology and life history of krill are unusual and puzzling. Recent studies have revealed that the krill genome (the amount of genetic information in each cell) is roughly twelve times the size of the human genome. We are not yet sure what the implications are, but it is an astonishing fact on its own. Krill can grow rapidly when food is abundant and shrink when food is in short supply. This is not simply a krill weight-loss exercise; krill grow and shrink by molting their shells every month or so, and when starved they must downsize with each molt. This trait complicates attempts to determine their age. We know from keeping krill in aquaria that they are long-lived, with life spans exceeding those of many of their predators. This means that at any point in time the population of krill is the product of many seasons, not just one or two, a discovery that has revolutionized the science of krill ecology and the management of its fishery.

*Sixth, krill are a delicious, nutritious food source.* The largest animals that have ever lived on our planet are blue whales, and the largest population of these giants used to feed exclusively in the Southern Ocean on one single food item: Antarctic krill. The population of blue whales was estimated at over one hundred thousand individuals before commercial whaling reduced their numbers by ninety-nine percent in the early 1900s.

Blue whales feed in Antarctic waters for only six months of the year; the other six months they migrate thousands of kilometers north to breed, during which time it is believed that they do not feed at all. How do they manage this? The answer must surely be in the abundance of krill and its nutritional value. Other baleen whale species, such as fin whales (the second-largest animal to have lived on Earth), humpback whales, and uncounted numbers of minke whales, all follow the same pattern of feeding in Antarctic waters and migration northward in winter.

In addition, krill are a prized food source for tens of millions of penguins, flying seabirds, seals, fish, and squid. How could this eco-system, this single krill species, have supported such a massive level of predation?

*Seventh, Antarctic krill have been the target of the largest fishery in the Southern Ocean for the past forty years.* Few people are aware that there are several active fisheries operating in the Antarctic region. The larg-est of these fisheries is for krill, and this activity has been going on for over forty years. After Antarctic seals and whales had been hunted to near extinction, the fishing industry turned its attention to krill. But after forty years of development, the industry still struggles to produce marketable krill products that justify the massive expense of fishing in the Southern Ocean. The fishery's slow development has allowed conservation efforts to expand, but the enigmatic nature of krill makes even their conservation difficult.

~

We know that krill are ecologically and economically important. We also acknowledge that they are fascinating, beautiful creatures in their own right. So why have they been misunderstood and misrepresented for so long? Living at the far end of the world under a layer of seawa-ter and ice proved a reasonable strategy for krill to keep a low profile for several millennia. However, once humans found the whale-rich Southern Ocean they began rendering it considerably less whale-rich, and they also began noticing the rather shy but utterly essential food of the giants—krill.

We humans have a tendency to see a resource and immediately as-sess whether we could use it to allow us to access a higher standard of living. This happened to krill in the 1960s, and since then considerable effort has been expended on catching krill and trying to commercially exploit krill as a resource, but this exercise has not been particularly successful, and as a species it has remained off the wider public's ra-dar—until now.

Krill are at a crossroads. As the Southern Ocean changes and opens to more human exploration and use, krill are once more under scru-tiny for their economic possibilities. Krill oil is now widely available in drug stores and supermarkets, and questions are being asked about the nature of krill and the sustainability of the fishery. This book is an attempt address these questions, but more than that. In the following pages, I hope to sing the song of krill so that a wider audience can appreciate their ecological importance and perhaps share the admiration some of us hold for this enigmatic, beautiful crustacean that has lived at the bottom of the world for millions of years. Hopefully, with a bit more understanding, they will continue to do so into the unforeseeable future.

*Chapter Two*

# Going with the Floes

The sea has never been friendly to man. At most it has been
the accomplice of human restlessness.
—Joseph Conrad, *The Rescue*

I can see the Southern Ocean from my kitchen window in the
south of Tasmania, Australia's island state. This vast tract of water stretches some three thousand unbroken kilometers (roughly
two thousand miles) from the local beach to the icy shores of Antarctica. In summer, icebreakers sail south past my windows on voyages that will take them well over a week to reach their destination.
If I am lucky, I can see the great ocean wanderers—whales and albatrosses—that are visiting Tasmania as part of their annual winter
migrations from polar waters.

The Southern Ocean is also home to Antarctic krill, but krill are
not found everywhere. Due south of Australia it would be many days'
sailing before krill would be first encountered. I have spent many
hours on the long sea journey from my home to the Antarctic wondering why krill are so abundant in some parts of this awe-inspiring
ocean and absent from others. Of course, the ocean is not a uniform
environment; great currents stir the waters, unseen thermal barriers

separate water masses, and, far to the south, ice advances and retreats in an annual rhythm, creating an underwater seascape like no other. This is the watery world that krill have evolved in, and their life cycle is tuned to both the constants and the seasonally changing elements of their environment.

The Southern Ocean is the stormiest body of water on Earth. It is covered in ice for half the year and is up to four kilometers (two and a half miles) deep, so it is not surprising that many of the animals that have made these waters their home have remained unknown until relatively recently. The oceans surrounding Antarctica were only first explored in the eighteenth century, so Antarctic krill were probably first seen by sealers or whalers when they cut open the bellies of their krill-eating quarry.

Krill rarely approach the shore, so studying krill in the Southern Ocean requires a capable vessel. Scientists often teamed up with commercial whaling expeditions whose sturdy vessels were built to withstand the punishing extremes of the Southern Ocean. In the 1980s scientists gained access to ice-breaking research vessels that allowed them to penetrate the habitat of krill at any time of year, which led to many breakthroughs in our understanding of krill. These vessels were also capable of catching krill, keeping them alive, and moving them to less hazardous environments where research on living krill could be carried out in controlled conditions. In a surprising turn, future scientific breakthroughs are likely to come from samples collected from commercial fishing vessels that can remain in the Southern Ocean year-round.

Antarctic krill are found only in the broad and turbulent band of water that surrounds the Antarctic continent. These restless waters at the bottom of the world are referred to as the Southern Ocean, but this designation is not without controversy. Geographers, oceanographers, and politicians debate the boundaries of the Southern Ocean, and some question its very existence. More conventional oceans, such as the Indian, Pacific, and Atlantic Oceans, are sandwiched between continents, and their borders are clearly defined. All three oceans, however, fade in their southern reaches and merge into the swirling waters that surround Antarctica. Even the southern boundary of the

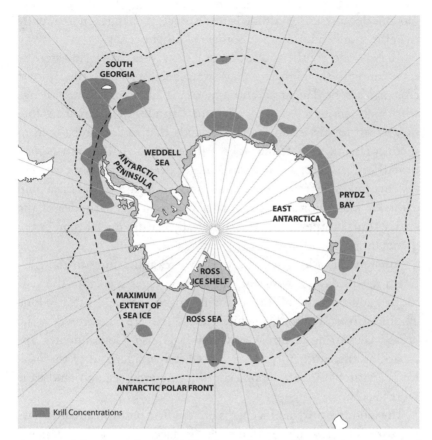

A map of Antarctica and the Southern Ocean showing the position of the Antarctic polar front (or Antarctic Convergence) and the average northern extent of sea ice in winter. The main areas where krill concentrate are also shown. (Drawing by Marcia Rackstraw)

Southern Ocean is not as distinct as one might imagine. The Antarctic continent squats under a massive layer of compacted snow and is surrounded by a seasonally frozen ocean. Where the ocean ends and the continent begins is not neatly distinct. The Antarctic continent itself rarely reaches the ocean in rocky or sandy shores. Instead, great ice shelves and glaciers flow off the sunken landmass and form the real southern physical boundary, the "great ice barrier" described by the earliest explorers who despaired of ever reaching solid continental ground. Because this coastline is made of ice, it changes more rapidly than the more enduring rock borders of other oceans.

To cross the Southern Ocean is still risky; it is the world's stormiest ocean with the highest average wave height, so even the shortest crossings can be dangerous and unpleasant. Ships must first traverse what mariners have referred to as the latitudes of the roaring forties, famed for the strong westerly winds that sailing ships used to speedily circumnavigate the globe. Farther south are the aptly named furious fifties, where a constant parade of deep low pressure weather systems progress from west to east and make life on board a ship distinctly uncomfortable. Eventually, in the higher latitudes of the sixties, things settle down atmospherically, but then ice becomes the dominant physical force.

Antarctica is a long way from anywhere, and sailing to the continent means days or weeks at sea. These voyages accustom the visitor to the immensity of the ocean, its movements, and its deep blue emptiness. Vistas of rolling, white-capped seas are punctuated by the occasional soaring albatross and very little else. Consequently, approaching the continent, or the vast fractal wonderland of the pack ice, the visitor begins to appreciate the emerging sea and landscape more keenly.

The looming towers of icebergs are the first break in the undulating two-dimensionality of the ocean surface. At first the bergs are sighted individually and distantly—eroded giants fast on the path to liquid oblivion. Icebergs become more common below latitude 60° south and, being younger, are more varied in size, shape, and color. Vast, flat-topped bergs the size of small countries vie for attention with the crenulated ramparts of crumbling ice castles.

The immense whiteness of the freshwater ice is perforated by the cobalt blue of cavities and chasms. Still more colorful are bergs shot through with vivid stripes of translucent turquoise, and the holy grail of the ice enthusiast—the deep dark green of the jade bergs. The bergs take on the color of the sky, changing constantly from piercing clarity at midday to subtle mauves and pinks during the timeless summer sunsets that merge into equally drawn out dawns. But all of this is merely the entrée, a taste of the wonders that the continent and its frozen fringe have to offer.

As the first barriers of ice and rock are reached, everything changes. Suddenly animal life is abundant; in the water, on the frozen ocean, on the few patches of exposed land, and in the skies above. Most Antarctic animals depend on the ocean for their existence. Although the open ocean is not sterile, the pack ice undergoes an annual cycle of melting and refreezing that renders the icy waters highly productive. In spring and early summer, pea-green blooms of microscopic plants explode on a planetary scale just behind, and even within, the receding pack ice. These prairies of the ocean are grazed by vast ocher swarms of krill.

Such an abundance of marine life has led to the evolution of an array of large animals that exploit this highly seasonal cornucopia. Great whales used to gather here in the millions, traveling from their breeding grounds in the balmy tropics to gorge for the few short months of the Antarctic summer on meals of krill that would sustain them on their long migration north and back again. Seals breed on the ice and exploit the food supply below and around them. Flying birds come from as far as the Arctic, and many breed in huge colonies on the small parts of the continent that become ice-free in the brief summertime. Penguins use the scattered rocky outcrops to raise their young in a frantically short period when the snow has melted and food is abundant. Thus, for the last hundred kilometers (sixty miles) as a ship approaches Antarctica, the ocean is suddenly alive with vibrant, noisy, and frenetic creatures making the most of the narrow window of opportunity to feed, breed, and hopefully survive another winter.

Although the pack ice, formed from frozen sea, is white on top, on the bottom it is heavily encrusted with complex communities of microbes. These ecosystems form on the underside of the floes and in their icy interior, which keeps them close to life-giving light. As the ice decays, the green, yellow, and brown of the living floes gradually emerge, and life itself begins to tint the physical world. The seals add daubs of color to the surface of the floes through their deposits of startlingly blood-red excrement, evidence of a krill-rich diet. Seals, like most of the larger animals of the pack, are rather drab, but their movements and the occasional flash of pink from their gaping jaws alleviate the tonal monotony.

But such monotony is not unusual—most marine mammals and birds the world over favor unexciting color schemes; the uniforms of gray, mottled beige, and contrasting black and white are common to whales, seals, and seabirds from the tropics to both poles. Onshore, the penguin colonies are more colorful, despite the near-monochromatic nature of their tuxedoed inhabitants. Sloppy pink guano formed from partially digested krill seeps into every crevice to the extent that these colonies are clearly visible from space—and can be smelled from many kilometers away. The nutrient-rich and pungent excrement fertilizes the surrounding area, resulting in swathes of pink, yellow, and green blooms of algae that colonize rocks, ice, snow, and ocean.

Penguin colonies are a riot of colors, smells, and sounds and are a vibrant counter to the stereotypical vision of Antarctica as a colorless, sterile wasteland serenaded by the white noise of the wind. To be sure, the animal life of Antarctica cannot rival the primary colors visible in the tropics; there are none of the species that bring color to more northern ecosystems: no reptiles, no amphibians, no butterflies, and no songbirds. But the fringe of Antarctica is far from a colorless sterile environment—at least in summer.

Moving inland from the living fringe of Antarctica, the signs of life—the colors, the sounds, the smells—diminish as the distance from the bountiful ocean increases. Fifty kilometers (thirty miles) inland, life diminishes in scale, with drab mosses and lichens eking out a silent existence in the ice-free valleys and on exposed mountain ridges. Eventually, even the bacteria and algae retreat into the interior of rocks, seeking protection from the life-threatening environment. The animals that dare to venture into the continent's ice-cloaked interior lose color and blend into their environment. Snow-white snow petrels roost in crevices in the mountains, and the coloration of the predatory skuas mimics that of the brown-gray rocks on which they breed. Two hundred kilometers toward the pole and multicellular life is essentially absent.

During autumn, most human and animal life deserts Antarctica, leaving a silent, sparsely inhabited, colorless continent. Paradoxically, the sole native resident during the long, dark winter is Antarctica's most colorful: the emperor penguin.

Towering above the Antarctic continent is a seemingly limitless dome of ice formed from the accumulation of eons of blizzards. Antarctica is bitterly cold, and the scant precipitation that falls is powdery snow. The snow slowly accumulates, and as the layers build up, their weight compresses the snow into ice. To the first-time visitor, the hard, slippery nature of the Antarctic ice is unexpected—a far cry from the soft snowdrifts of the temperate zones. This accumulation of continental ice can build up to remarkable depths—over four kilometers (two and a half miles) in places—and over a million years of snowfall can be found beneath the wind-scoured surface of the icecap.

But the ice does not remain stationary. Inexorably, it flows toward the northern coastline in slow-moving crystalline rivers. This flow occurs in great crevassed ice sheets, thousands of kilometers across, and also in more discrete, faster-flowing glaciers that find their way between protruding mountain ranges and down hidden valleys to the distant sea. This ice is pure, fresh water formed only from the sparse continental snowfalls.

The Antarctic ice cap contains approximately sixty percent of the planet's freshwater locked up in its frozen, slowly moving heart. Pure ice is less dense than seawater so when, after millennia the glaciers and ice sheets reach the ocean, they float on the dense, saltier, warmer water. The sheer mass of ice behind continues to push the glaciers onto the ocean's surface, and eventually the warm ocean and the rhythmic movements of the ocean's tides and storms begin to crack and break the brittle glacial tongues. Icebergs are formed as the ocean erodes the protruding tongues of ice and huge chunks of glacial ice are cast adrift on the ocean. The ice sheets also float, forming vast cliffs of ice that snake along the coastline for hundreds of kilometers, spewing tabular icebergs—the size of small countries—when they crack and break.

In contrast to the ancient continental ice, sea ice is ephemeral; most is formed and is gone within a year. In the Antarctic autumn, as daylength decreases, the air temperature plummets and the surface of the ocean begins to freeze. Glaciologists have developed nearly twenty different names for the various stages of ice formation, but a small number of terms describes the important processes. *Frazil ice* consists of fine needles of ice suspended in a watery matrix and is a precursor to more structured development. When a thin layer of ice crystals glazes over the surface, making the ocean surface appear as a mat and unreflective, it is called *grease ice*. Wind and wave action stir up the forming ice into flat semicircular structures called, not surprisingly, *pancake ice*, which combine and fuse, eventually forming the more rigid and expansive pack ice.

The pack can grow to over two meters thick in a single year and can be far deeper in places where the incessant movement of the ocean rafts huge blocks of ice, one on top of the other. The surface of pack ice may look smooth and flat, but below the surface it is a labyrinth of ridges, caverns, and inverted ice mountains, and we are beginning to learn that this upside-down wilderness is the winter playground of juvenile krill.

Sea ice is formed by cooling from the atmosphere above, and as it grows and thickens, the floating ice begins to seal the ocean from the atmosphere. This surface layer of ice insulates the relatively warm ocean from the much colder air, but because it is white it simultaneously reflects the little sunshine and heat reaching the Southern Ocean in winter. This balance dictates the rate of ice formation. Because of the vast volume of the ocean, which is much warmer than the atmosphere, the colder winter temperatures only directly affect the surface layer.

But the cooling does produce other, less obvious outcomes. Forming ice ejects salt as it freezes. This makes the sea ice lighter, and the layer of water below the forming ice simultaneously becomes both saltier and colder, hence denser. This dense water sinks to the ocean depths, flowing north along the seafloor and initiating a process that drives much of the circulation of the planet's oceans. Sinking water must be replaced by a flow of water elsewhere, effectively turning the interior of the Southern Ocean into vast, slow-moving rivers of water of different densities flowing north and south, carrying oxygen, nutrients, and carbon dioxide across the planet.

The pack ice builds steadily during autumn, extending northward to cover some twenty-two million square kilometers (eight and a half million square miles)—essentially doubling the area of the Antarctic continent—by the end of winter. Some of the ice is firmly attached to the shoreline of the continent and is known as *fast ice*, but the most ice forms a seemingly endless white blanket covering the surface of the restless ocean. The tides still raise the ice-covered ocean surface twice a day, and these movements cause the pack to crack and fracture. Beneath the ice the great ocean currents continue to stir. Some of these currents are driven by wind and some are driven by density differences in the ocean interior, but these gears of the great ocean machine continue to spin, even when the surface of the ocean is covered in ice. The ice layer is never more than a few meters thick so surface processes do not get in the way of the business of moving the planet's water around.

The pack ice swirls in patterns driven by the underlying currents and by the wind—nothing is static. There are great gyres that surround the Antarctic continent, and the ice swirls around, following the lead of the underlying water. These gyres affect ocean physics, and they have driven the way in which life has evolved in the Antarctic region. An Adélie penguin can step onto the ice from its land-based colony in autumn and be swept to the west on the southern limb of a gyre, before being transported northward then eastward on the gyrating ice, finally being delivered back to its colony just in time for spring. This glacial merry-go-round also keeps the penguins in constant contact with the teeming swarms of krill that overwinter beneath their feet.

The sea ice zone of the Southern Ocean is not just a flat expanse; it is a dynamic environment that can change rapidly. Ice generally grows outward from the continent's cold heart toward the warmer ocean to the north. This growth is not uniform because it is affected by regional winds and currents. In some areas conditions favor rapid ice growth, in others the growth is slower, and in still others newly

formed ice is rapidly whisked away by winds and currents. In polynyas, which are areas of open water in sea ice, the fierce, bitterly cold,
katabatic winds from the interior of the Antarctic continent blow the
forming ice northward, leaving patches of open water in its wake. This
encourages more ice to form on the newly exposed ocean; polynyas
work like ice factories, making ice and exporting it to the north. The
animals that live in, on, and under the sea ice have evolved in this
ever-changing winter environment.

Sea ice also has a unique ecosystem and, as in most ecosystems, life
begins with plants. Photosynthesis in the open ocean can only occur
in the well-lit surface layer of the ocean, generally less than two hundred meters (roughly 219 yards) in depth. In aquatic systems, there is
a physical tug-of-war; light comes in from the top, but gravity acts
on particulate matter from the bottom. For the single-celled phytoplankton, which are responsible for photosynthesis in the ocean,
accessing sunlight is an ongoing struggle.

The phytoplankton, which comprise species of algae and other
microorganisms, have limited swimming ability, so they tend to sink
away from the sunlit surface of the ocean. Even when these plants
can maintain themselves in the sunlit zone, they continuously lose
access to essential nutrients as nutrient-rich cells die or are eaten and
converted to particulate matter, which sinks to the deep ocean. Surface waters are thus stripped of nutrients over time, and these must
be replenished for plant growth to continue.

The surface zone gains nutrients from the atmosphere as dissolved
gases or dust from land via glaciers and erosion and from the sedi-
ments on the seafloor. Nutrients also reemerge in the surface waters
as deep water, rich in nutrients from the decay of organic matter that
has sunk from the surface, upwells. Upwelling zones are among the
most productive ecosystems on the planet, and Antarctica is ringed
by a series of these inverted waterfalls that bring nutrient-laden deep
waters to the surface. The productivity of the ocean is dependent on
the balance between nutrient gains and losses, and the efficiency of
nutrient recycling in the sunlit zone.

The seasonal cycle of productivity in the open ocean is a slave to
the varying light regime and to the physical changes in the water that

occur seasonally. The surface water is warmed in spring, separating it from deeper, denser cold water by a density barrier known as a thermocline. In summer, the winds mix the surface waters and the plant communities above the thermocline, which keeps the algae within the sunlit zone where they can photosynthesize. In winter, as the sun grows weaker and the winds stronger, the thermocline breaks down and the surface waters mix to much greater depths, bringing essential nutrients to the surface but restricting the growth of algae, which spend much of their time away from the feeble winter sunshine.

The seasonal cycle in polar oceans differs slightly from that in lower latitudes because of the presence of ice and the extreme seasonality. Plants must cope with twenty-four hours of darkness in winter and twenty-four hours of sunlight in summer. In winter, the lack of sunlight means there is little photosynthesis within the open ocean. Stormy conditions mix the surface ocean to greater depths, exacerbating the problem of low light levels. Coupled with these factors, the thick cover of surface pack ice further reduces the amount of light reaching the ocean's interior. The thicker the ice and the greater the snow cover the less light gets through, which drastically reduces the depth at which photosynthesis can occur.

Ice, however, offers a substrate on which algae can grow. Because sea ice floats, these ice algal communities are kept close to the surface of the ocean, allowing them to be in an optimal position for the meager solar radiation available. These communities incorporate into the forming ice, growing over autumn and winter. An icebreaker punching its way through pack ice in late winter or spring reveals a dirty yellow-brown underside hiding under the pristine white surface of the ice floes. These upside-down pastures are often the grazing places for herbivores, such as krill, which can be seen wriggling uncomfortably in the sunlight as their world is overturned by the violent passage of the ship. The pack ice is one of the few productive oceanic zones during the darker months, and these under-ice communities are oases that krill, particularly in the larval stage, rely on.

Those who have been lucky enough to venture under the layer of pack ice in the spring have seen krill cruising, upside-down, on the underside of the floes using their feeding limbs to scrape the film of algae off the ice. The convoluted underwater environment of the pack ice, with its ridges, rifts, and leads, provides refuges for krill where they can hide from the few predators that overwinter in the Southern Ocean. The few diving expeditions that have visited the pack ice zone in winter have described a magical three-dimensional inverted ice-scape with millions of young krill migrating through the convoluted, icy network of ridges, caves, and plains.

For larger animals, the sea ice in spring provides a surface on which they can breed and feed, by diving through leads or through breathing holes that they actively maintain. Most krill predators, such as the large baleen whales and penguins, flee beyond the pack ice in winter, but some remain. These hardy species include the hugely abundant crabeater (in truth, krill-eater) seal, and the emperor penguin, which breeds on the fast ice in the depths of the coldest winters on the planet. Crabeaters make their home on the ice above the regions where krill are most abundant, making frequent forays into the underlying water to feed and then resting and breeding on the surface of the ice until spring arrives.

In spring, as the sun starts to bring some warmth and light to the Antarctic region, and as the days lengthen, the ice starts to change. Melting begins, first at the edges and then in the interior of the ice. Ponds of meltwater form on the once-white surface and are colonized by algae, turning the water dark, absorbing more heat, and thus accelerating the melting. In the reverse of the autumn freeze, the spring thaw starts in the north, and the ice edge moves south as it melts and is broken up by the waves and tides from the open ocean. The sea ice melts into water that is less salty than the encroaching seawater, so a stable layer of fresher water is left behind the retreating ice front. This layer provides a habitat in which algae can grow and stay close to the surface.

The fresher water also contains algae and nutrients released from the melting ice, and these begin to bloom in the favorable conditions. Satellite images of the Southern Ocean spring ice melt show huge fields of algal blooms in the wake of the retreating ice. These are a predictable food source for the herbivores, like krill, that have been living beneath the ice over the winter. In some years, these blooms are enormous and provide a bumper crop to fuel the krill reproductive season. In other years, the bloom is less extensive, or is at the wrong time, which leads to a poor breeding season for krill that year.

In the past two decades, satellite imagery of sea ice extent has become widely available. These images have allowed scientists to compare the regional amount of sea ice to the size of the krill population as measured from research vessels. In some areas, the relationship appears strong: years of extensive sea ice were followed by years with larger populations of krill, and years of less extensive sea ice resulted in fewer krill. This tidy relationship was simple to explain from observations of krill feeding on ice algae. If ice algae are the major food source for overwintering krill, then more sea ice in winter would mean larger algal pastures in winter, and greater algal biomass in the sea ice meant more phytoplankton would be ejected into the water on melting in spring. In short, more sea ice made for better feeding conditions.

Unfortunately, as with many studies, the collection of more years of data has converted a simple story into a much more complex one. Winter sea ice extent is no longer viewed as the sole determinant of krill abundance. We are learning that conditions for algal growth depend on more than sea ice area, and that these conditions have a regional component.

The pioneers of krill research, working back in the 1930s, noticed that the summer distribution of krill corresponded quite well with the winter distribution of sea ice. However, the relationship between distribution of krill and winter sea ice is complex, and striking exceptions to this relationship exist. For example, a large population of krill can be found in the waters around the sub-Antarctic islands of South Georgia and Bouvet Island, both beyond the reach of winter pack ice. In the ice-bound bays and coastal seas that penetrate the farthest south into the continent, Antarctic krill are scarce, which means that the presence of ice alone cannot predict preferred krill habitat.

Sea ice is thus only one of the features of the marine habitat that affects where krill can be found. Although krill are usually found swimming freely in the open ocean, it appears that features on the ocean floor also affect their distribution. Off the long linear coast of

East Antarctica, krill swarms are most commonly found just offshore of the continental shelf break. As a ship sails toward the Antarctic continent the echosounder shows the increasing abundance of krill as the ocean begins to shallow, then at the point where the continental shelf begins the krill disappear. On the continental shelf, another species of krill, ice krill, appears; the transition between species can be abrupt. This lack of Antarctic krill on the continental shelf off East Antarctica has been well documented by scientific studies, by the location of fishery catches, and by studying where penguins go to feed.

In contrast, off the Western Antarctic Peninsula krill can be found on the continental shelf and even extending into the intertidal zone. On my first visit to the Antarctic Peninsula a shipmate pointed out Antarctic krill swimming among fronds of kelp. I had just given an onboard lecture in which I claimed that krill were oceanic creatures unlikely to be found in shallow water. Yet there they were, clearly visible from a zodiac and as close to the shore as is possible. My experience was from the East Antarctic, where krill are usually found tens to hundreds of kilometers offshore. I am still not sure why krill exhibit such different patterns of distribution on opposite sides of the continent, but I am now always sure to emphasize the fickleness of krill when describing their distribution.

There are no continents at the northern extent of krill distribution so the features that limit the habitat in lower latitudes are more difficult to pick, but like the southern extent, it varies considerably around the continent. Off Eastern Antarctica, krill are rarely found north of latitude 60° south, whereas in the South Atlantic large portions of the population extend as far north as latitude 50° south. Such variation is likely driven by the way in which the patterns of ocean circulation affect the habitat of krill, but it demonstrates how flexible krill are, being able to flourish in many different environments.

What we do know is that krill prefer colder water. It is tempting to think of krill living in an extreme environment, but they actually inhabit a relatively stable world. The temperature of the water in the Southern Ocean varies little. It never heats to above 5°C (41°F) and, if it gets any colder than minus 1.8°C (−28.8°F), it becomes ice.

So thermally it is a very stable environment—most temperate and tropical marine animals survive in waters where they can experience temperature changes of greater than 20°C (68°F). Out of the water in the Antarctic, animals must survive winters where temperatures drop below minus 50°C (−58°F) and summers where sometimes the air temperature can reach 10°C (50°F). At their northern limit, krill are living in water that is always warmer than those living farther south. In the future, it is likely that this differential will increase as the north warms more quickly than the south, which is buffered by all that ice.

The circulation of the Southern Ocean at first glance seems straightforward. The world's largest current, the Antarctic Circumpolar Current (ACC), occupies most of the ocean area, transporting a mind-boggling one hundred and fifty million cubic meters (roughly one hundred ninety-six million cubic yards) of seawater a second at speeds of up to four kilometers (two and a half miles) per hour. This current, known to mariners as the West Wind Drift, is driven by the constant winds that circulate around the globe unchallenged by any landmasses at these latitudes. More subtly, the current is also a product of the density differences between water masses, and these forces produce this seemingly unstoppable flood of water. A few island groups and underwater ridges alter its flow pattern, but for most of the Southern Ocean north of latitude 60° south, the flow is from west to east.

Closer to the continent things begin to change. At the shelf break, where the seafloor rises from the abyssal depths to meet the shallower water of the continental shelf, another current takes over, and this one moves in the opposite direction to the ACC. This current is known as the Antarctic Coastal Current (or East Wind Drift), and although less voluminous than the ACC it can flow much faster in narrow jets. Unlike the ACC, however, the coastal current is not entirely continuous. Steered by features on the coastline and on the ocean floor, the coastal current veers northward in places, and is then guided eastward by the ACC before heading south and then west again, forming the series of clockwise gyres that affect the distribution of sea ice in winter. These current systems contain seawater with slightly different physical properties and thus become separated by a series of oceanic barriers or fronts. This is of immense importance to the animals that live in the region.

At the northern limit of the ACC there is a staggered series of oceanographic fronts. These culminate in the north at the Antarctic Convergence, the major oceanographic feature that most oceanographers recognize as the barrier that separates the Southern Ocean from the more temperate waters to the north. Here, the cold Antarctic waters sink beneath warmer sub-Antarctic waters, and this causes upwelling and mixing—the prerequisite for biological productivity. Cross this oceanic barrier in a ship heading south and the difference is often palpable. The air temperature drops and the humidity decreases—some people even swear they can smell the change.

The waters surrounding the Convergence are often hotspots for observing wildlife as they access the rich food supplies made possible by the increase in nutrients. Krill are almost never found north of the Convergence; for this reason, the position of the Convergence was chosen as the northern limit of the area managed by the convention that was developed to manage the krill fishery (see chapter seven). After decades of study we now know that krill are rarely found this far north—the latitude of the Convergence is usually in the fifties latitudes. Only near South Georgia, where the topography forces the ACC north to squeeze it through the Drake Passage between South America and the Antarctic Peninsula, does the distribution of krill approach the Convergence.

So what is the northern limit of krill distribution if it is not the Convergence? We still don't know. Despite the ACC being a single broad current, it too has considerable oceanographic structure. A series of east–west frontal systems divide the flow into horizontal layers, and the more the current system is investigated the more complex this segmentation becomes. One of these frontal systems may well prove to be the barrier that krill will not cross, but we are not yet ready to definitively identify the culprit.

All this talk of boundaries gives the impression that there are hard and fast lines drawn in the ocean and that animals strictly obey these boundaries. Not surprisingly, this is not the case. These oceanic

barriers between the water masses and communities have been moving as part of the response to the increased amount of heat in the
oceans. Because the distribution of krill has been linked to the positions of the currents and fronts, the unfortunate krill population will
face yet another set of challenges.

~

Detailed study of the ocean indicates that the general flow actually
consists of a massive number of embedded eddies, each spinning
within the overall flow. At boundaries, some of these eddies spin off
and can escape into adjacent water masses, and in the process carry
resident krill out of their normal habitat.

There are numerous reports of krill being found outside the boundaries of the Southern Ocean, around some of the sub-Antarctic islands that are generally thought to be to the north of their preferred
habitat, including one record from a Chilean fjord. There is even one
intriguing report of Antarctic krill being found in the stomach of a
humpback whale that was killed off the coast of tropical Queensland
in the 1970s. The drawings of these krill look convincing but how
could they have made their way to the tropics? One unlikely explanation is that the whale in question had a digestive disorder so the krill
remained undigested throughout the long seven-thousand-kilometer
(4,350-mile) journey to Australia from the Antarctic. Another explanation is that krill could survive in deep, cold ocean currents that
sweep north from the Antarctic and that the whale was able to find
them in this unlikely location. Antarctic krill prefer colder water, so
there would be no thermal barrier to them heading north at depth.
On the surface, in northern waters, they would rapidly encounter
water above 5°C (41°F), an increasingly uncomfortable temperature
for this species of krill. Off Queensland, the surface temperature is
around 25°C (77°F)—warm enough to cook a krill, but at two thousand meters (roughly 2,200 yards), it might be a rather comfortable
4°C (39°F) so theoretically there is nothing to stop them colonizing
the whole ocean at depth. But animal distributions respond to more
than just seawater temperature. We will probably never know how
those krill reached this far northern outpost, but it does highlight the

issue that barriers to species movement are not easily delimited and may vary with depth and time.

After decades of study we have a reasonable idea of where krill are to be found in highest abundance (see earlier map showing krill distribution). Generally, within their habitat, krill are associated with physical features: ice edges, island groups, the shelf break, and the edges of currents, or at certain frontal systems. Encountering large aggregations of krill out in the featureless open ocean is not impossible, but it is rare. The distribution of krill fishing vessels is telling; they concentrate on fishing in underwater canyons, in the lee of islands, and near shelf breaks. Fishermen choose fishing grounds for a host of other reasons as well that often have nothing to do with the overall distribution of a species. When they find a concentration of krill, they tend to stick with it, and it does not matter how many krill are in the vicinity. Other krill predators tell a similar story of feeding in predictable locations associated with physical features. This preference for regions where the seafloor is complex is a feature of many other species of krill too. Thus, when a krill aggregation is located in two dimensions the problem becomes one of determining the depth at which they are living. This is not as easy as it might seem.

The first real indication that krill were so abundant in Antarctic waters came from observations of extensive surface swarms that turned the water red. Captain Cook, on a visit to the South Sandwich Islands in 1775, was startled by "uncommonly white water" and the ship was diverted because the officer of the watch thought the water was shoaling. Cook later thought it must have been a school of fish, but in those waters it was more likely to have been a krill swarm. One whaler's account from 1892 exclaimed, "Whales' backs and blasts were seen at close intervals from quite near the ship, and horizon to horizon ... the sea was swarming with *Euphausia*." The great collation of krill research from the Discovery Expeditions in the 1930s consistently referred to the observations of surface patches of krill. Even the earliest attempts to fish for krill suggested using model airplanes or satellites to spot the red discoloration signifying the presence of krill.

Vibrant red surface swarms of North Atlantic krill were my introduction to marine biology, so in the 1980s I was keen to visit the

Antarctic and see this phenomenon expressed on a larger scale. Sadly, in all my voyages to the Antarctic, I have rarely seen surface swarms. Occasionally I have seen small isolated patches of discolored water that turned out to be krill, but never have I seen the extensive ocher fields of krill written about in early reports. Few of my colleagues who have worked with krill extensively since the 1970s have seen a surface swarm of Antarctic krill. Krill undoubtedly used to swarm at the surface in the early years of the twentieth century, but such swarms appear to be less common now—or they are less reported. This is one of the great mysteries of krill biology. Perhaps modern scientists spend too much time examining their computer screens rather than scanning the water. But if krill have really altered their depth preference, what could have provoked such a drastic change in behavior? Some attribute this to the disappearance of their most voracious predator—the great whales. Some to the opening of the ozone hole that has driven krill deeper to avoid UVB radiation. Others remain skeptical that surface swarming ever happened in the first place.

Because of these early observations, our studies on krill have focused on the upper ocean. This concentration of effort is not just based on historical knowledge, it is also a reflection of our ability to study krill—most of our scientific sampling gears really only works properly in the top two hundred meters (220 yards) of water. I will expand on this in the next two chapters, but it means that most of our knowledge about the biology and life cycle of krill comes from this restricted surface zone. What if krill lived deeper? How would we know? How would it affect our understanding of krill?

Unmistakable images of individual krill or krill swarms at depths far deeper than two hundred meters (220 yards) and even deep under ice shelves are emerging from the recent deployment of underwater cameras. One study even found adult krill near the seafloor at a depth of four and a half kilometers (roughly three miles). In recent years, the krill fishery has regularly targeted aggregations far deeper than two hundred meters (220 yards). We have known for over fifty years that krill eggs and larvae can be found at great depths, but it has always been assumed that adult krill were creatures of the sunlit zone—this, after all, is where their food is found. These recent findings of deep-living krill are likely to change the way in which we look at krill and our assumptions about how they might be able to adapt to future changes.

The remote Southern Ocean, like the rest of the planet, is experiencing rapid change. The ocean is getting stormier and becoming warmer and more acidic. Sea ice is becoming less extensive in summer, and, in some areas, is now absent in winter. Sea ice retreat starts earlier in the spring and advances later in autumn than in the recent past. Faced with such significant changes to their habitat, animals like krill have a number of options: adapt, move, or decline. It is unclear which combinations of these options the krill population will take, but there is more to predicting the future than examining the physical symptoms of change. The physical environment defines where krill might be able to live, but it does not indicate exactly where krill can be found, now or in the future. The boundaries of ice and temperature, seafloor and surface merely delineate the possible habitat of krill. The exact location of krill swarms is the result of interactions between the krill and their watery milieu, and these interactions occur on an evolutionary timescale as well as much more immediately through their individual and collective behavior.

For nearly a century we have known that krill migrate vertically, and this mass movement of the krill population occurs daily. During the day, krill lurk in the darkness at depth, whereas they move toward the surface at dusk, descending again at dawn. This behavior avoids the attentions of visual predators; a similar behavior is exhibited by a wide range of animals throughout the ocean. Imagine the mass movement of trillions of krill from the ocean's depths to the surface, stirring up the water, eating whatever is in their path, and leaving a rain of detritus, cast shells, and waste material in their wake. This is a phenomenon of global scale, and this happens every day—or almost.

Close examination reveals that not all individuals in the population are migrating simultaneously; some may remain at depth and others may migrate upward, feed, and then sink downward when satiated. Still more may remain in the surface layer or at depth depending on the season. There are also differences in the behavior of life-history stages, with younger animals migrating to a different rhythm than adults. So, in a familiar fashion, krill evade attempts to explain their behavior in simple terms.

There is now growing evidence that krill migrate horizontally as well as vertically. We know that egg-laying females are found farther offshore than the rest of the adult population, and that juveniles are often sensibly found in areas distinct from their voracious parents. These patterns suggest that krill are able to migrate between areas. The entire population off the well-studied Antarctic Peninsula appears to migrate, from offshore waters in summer across the continental shelf into the shelter of bays and coastal underwater canyons in winter—several hundred kilometers. How can krill sustain the swimming speeds necessary, and, more importantly, how do they navigate? This migration pattern has only been revealed in recent years because the pack ice in the Antarctic Peninsula region has shrunk, making it possible for research vessels to enter the area in winter. We should not be surprised that animals like krill are capable of long-distance migrations. Terrestrial invertebrates, such as the monarch butterfly, are known to migrate across whole continents, and marine species smaller than krill, such as eel larvae, cross whole oceans with pinpoint navigation skills. How krill migrate, and even whether they do, is still a matter of heated debate whenever krill biologists get together.

The evidence for krill migration clashes with the traditional view of krill that their distribution is largely determined by current flows. Assuming currents are the dominant influence is understandable given that krill live in and around the world's largest current system. Since the dawn of krill research, it has been assumed that the large population of krill on the South Georgia shelf had come from elsewhere. And so was born the concept of krill flux. There is, it is supposed, a river of krill, a "conveyor belt" of sorts, carrying young and adult krill northward from the Antarctic Peninsula to South Georgia. This concept was particularly attractive for those who wished to fish for krill at South Georgia, because it implied that all the krill were being constantly drifted past the island and from there into the open ocean where they would perish. Scientists from the Soviet Union argued that, because any krill that got past the island would be lost into the oceanic void, it would be possible, even responsible, to harvest large quantities of krill in the waters around South Georgia without damaging the ecosystem.

The concept of krill flux obviously has implications for the way the krill fishery is managed and has remained in currency for the past thirty years despite limited supporting evidence. Part of the reason for this persistence is that it is a difficult problem to address. How do you track individual krill or krill swarms across the ocean? A solution to this conundrum has been to simulate the ocean currents of the South Atlantic in a computer model and to drop imaginary particles into the flow field at various places to see where they might end up. Krill rarely act like particles so the results of these studies to date have been equivocal, and the push is to improve them to include more krill behavior and more detailed structure of the ocean currents. Who knows where the simulated krill may end up?

If krill flux is only a minor determinant of krill distribution and most krill remain in an area, then it ought to be possible to distinguish krill stocks using differences in their genetic makeup. Molecular techniques seemed the ideal way to address the question about whether krill populations are just drifting around the Antarctic continent on ocean currents (one genetic stock) or whether there are distinct populations (separate stocks) in the different regions of the Southern Ocean. Numerous studies have looked for such markers of distinct krill populations, but none have shown any. However, this does not mean that these distinct populations don't exist. These studies also revealed a very high degree of genetic diversity in samples from a small area, which would make it extremely difficult to detect differences between areas if they existed. The recent discovery that the krill genome is massive—twelve times the size of the human genome—means that using genetic markers to distinguish krill populations is a distant dream.

But help has come from an unusual source. Pollutants have been detected in the Southern Ocean and include all of those found in more temperate waters, including radionuclides from bomb tests; persistent chlorinated pollutants, such as PCBs; heavy metals; and flame retardants. These chemicals arrive from the atmosphere and in the ocean currents—and some are even derived from human activities in Antarctica. Animals pick up a signature of pollutants that reflects the oceanic environment where they have grown up. Studies show that adjacent populations of krill have very distinct signatures, suggesting that these populations have remained separate for a considerable period of time. This finding suggests that using such compounds as markers of mixing between local populations might be possible, and we might soon be able to determine the rate at which krill move between areas around the Antarctic and the degree to which this movement is passive or active. But could krill actively maintain themselves in an area when their swimming speed is slower than the average current velocity?

Recent advances in the method oceanographers use to measure currents have revealed a complex, underwater world with contrasting water flows layered one on top of the other. So, the average current velocity does not tell the whole story. In the horizontal plane, there is a similar complexity, with small distances separating divergent water flows. Computer models, like the ones used to simulate krill flux, often fail to incorporate this complexity, and they smooth out the currents into long-term averages. But to an animal that is capable of fast swimming and navigating, this three-dimensional fluid maze offers the possibility of staying in one place or migrating, merely by making small vertical or horizontal movements. Krill swarms are selective; they aggregate in areas where current flows are low and avoid the high-speed jets of water. So krill, like schooling fish, can potentially remain in favorable areas and avoid less desirable locations through a combination of behavior and canny exploitation of the current systems. Whether this is sufficient to account for long-distance migrations is a subject of active study.

I will examine some of these aspects of krill behavior in chapter 3, but for now it is enough to understand that krill distribution is highly patchy on all spatial and temporal scales. There are some areas, at certain times of year, where krill are highly abundant, and this abundance is relied on by predators, krill researchers, and the krill fishery. There are other areas where krill aggregations are perennially scarce.

The final point to make is that our generalizations about where we might find krill always have exceptions—we are, after all, dealing with an animal that has some degree of control over where it lives, and we are not able to understand exactly what drives its behavioral choices.

*Chapter Three*

# Labors of Love

Much of our oceanographic sampling gear ... was recognized
as inefficient over half a century ago and it is becoming
increasingly obvious that in our approach to any specific
ecological problem old methods must be cast aside,
giving way to the revolutionary or new.
—James Marr, *The Natural History and Geography of the
Antarctic Krill*

F ew people have experienced the hardship, tedium, and excitement of long oceanographic research voyages to distant waters. Prior to my first foray into Antarctic waters, my longest stint at sea was a mere three weeks, the amount of time it takes just to get my sea legs. The real work starts in the following month or two. These long voyages are intense periods of scientific activity, but they are also a test of our ability to work for long hours in close proximity with a small, but disparate, group of people in a workplace that wallows constantly. Achieving our scientific goals requires considerable social skills, but it is also a powerful bonding experience. In the best cases, lifelong friendships are born. In the worst case, implacable enmities evolve. But, at sea, the focus is always on completing the mission; it is usually only when the ship has docked, and the debriefs with friends and family commence, that true emotions can come to the surface. Nobody forgets a long Antarctic voyage.

Open-ocean krill research is an industrial-scale exercise involving large ice-breaking vessels, unwieldy pieces of sampling gear, teams of scientists, technicians, and the ship's crew—and, because krill are always distant from the port of departure, months at sea. Just getting a sample of krill onto the trawl deck is a major operation, and we can repeat this a couple of hundred times on a research voyage. Once we have decided to trawl, the ship slows, the trawling team is assembled, and the sampling net—usually a rectangular device with a cumbersome eight-square-meter opening—is prepared. The cod-end is attached to capture the krill we hope to collect, and the ship's crew, with their hard hats, safety lines, and floatation vests, take over. The gantry system squeals in mechanical pain as the net rises overhead and backward, then is slowly lowered into the ocean. When the net reaches a depth of about ten meters, the mouth is electronically opened and it sinks slowly into the depths, devouring the animal life in its path. At our chosen depth, the winch engages, and we slowly haul the net back to the surface, where it is closed and then hoisted back onto the deck. Then the scientific team springs into action. The net is washed down with streams of icy water, and the cod-end, with its precious cargo of animal life, is carefully removed and inspected. Sometimes the catch is so large that the krill are backed up into the net beyond the cod-end. Sometimes there is little to show for thirty minutes of trawling, and only close microscopic examination reveals the nature of the catch. The sample is then whisked away to the adjacent laboratory so that it can be examined, subsampled, measured, and commented on by all who were involved in the trawling operation. This is where scientists and crew gather eagerly to see what bounty the ocean has returned.

The trawling operation is usually performed on a slippery, heaving deck accompanied by a cacophony of mechanical sounds emanating from the ship's winches, hydraulics, and engines. It is never quiet on a research vessel, and life becomes attuned to the unique sounds of each operation. The damp and rusty deck throbs to the rhythm of the

Retrieving a large (eight-square-meter [nine-and-a-half-square-yard] mouth opening) scientific net of the type used to catch krill off the stern of the research icebreaker *Aurora Australis*. (Photo by author)

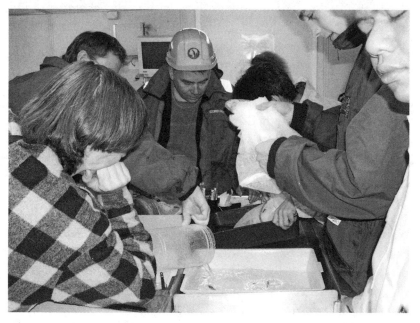

Inspecting the catch in the laboratory on the research icebreaker *Aurora Australis*. A younger incarnation of the author is in the middle, wearing a hard hat. (Photo by author)

ships propellers as the vessel makes slow headway into the swell. Occasionally a rogue wave will find its way up the stern ramp and onto the deck, causing chaos with the assembled equipment and operators.

Sampling continues day and night so there are usually two shifts of the scientific team and ship's crew, and a spirit of friendly competition develops between the day shift and the night shift. What comes up in the net is always a surprise. Each shift vies for the largest catches or the most unusual captures, and the tally for each shift is proudly displayed on the whiteboard in the ship's mess. This sampling exercise is repeated dozens of times on every research voyage, and, although it is cold, wet, noisy business, it always invokes the spirit of discovery.

Alongside the blunt instruments on the trawl deck, the ship's electronics work in the background, measuring ocean variables and detecting krill. These sampling voyages are high stakes—ship time is expensive and in short supply so there is pressure to utilize every minute available in the month or two that the vessel is down south. But of course, the serious work goes on back in the laboratories and institutes where the krill are measured, the data are compiled, and the papers are written, and this phase takes years, rather than months. This exercise has remained essentially unchanged for the last hundred years, although these days it involves more electronics and safety gear. I often step off long voyages wondering why there has been so little development in the methods we use to sample the ocean.

For scientists who work in the open ocean—the largest biome on the planet—spending time immersed in ocean ecosystems is a surprisingly rare event. Most of the ocean is deep, cold, and far from land, and the deepest reaches of that fragile band where life occurs are beyond the range of scuba divers. The water pressure alone in these twilight or perpetually dark waters is so extreme that unaided human exploration is difficult, expensive, and risky, so most research takes place from ships on the surface with mechanical or electronic assistance: sonars, nets, submersibles, and, increasingly, robotic vehicles equipped with advanced sensors. This separation between scientists and the animals we study means that our impressions of marine life are nearly always secondhand; data are collected remotely without the careful observations that are possible in terrestrial systems. There are centuries of accumulated information and observations on terrestrial natural history collected by scientists and amateur naturalists. This is rarely the case for marine organisms, particularly those that are found far out in the ocean, like Antarctic krill. There are exceptions. Sometimes krill venture into human-accessible waters, as they did in the Bay of Fundy during my early years as a scientist. Those normally deep-living North Atlantic krill came to the surface in swarms, providing a rare opportunity to see and sample the ocean's inhabitants at the same time.

Whether in the Bay of Fundy or in the Antarctic, our understanding of krill is piecemeal at best; our ability to spend time with krill is highly limited, and, until recently, our sampling methods have been crude. Consequently, there are large and important gaps in the science that need to be filled if we are to understand the ecosystems where krill play such an important role. But it is in the Antarctic that the need for answers is most critical, where climate change and fishing have the greatest potential for impact on an ecosystem that is still only poorly understood. For scientists who study krill, the challenges are formidable. It is difficult and costly to get to the Antarctic, and opportunities to study krill are rare.

For those fortunate research expeditions that gain formal approval and make it to the Southern Ocean, more obstacles await them. Tools for studying krill in their natural environment are few. It is difficult, although not impossible, to experiment with live krill on board a constantly moving research vessel. Transporting them from the Antarctic to laboratories back home is logistically challenging but provides a more controlled environment for study. Historically, most krill research has not involved living krill but has relied on studying dead krill under microscopes. This absence of research on krill in their natural habitat means that we have overlooked their ecological complexity, including their interactions with their environment and with others of their own kind. We have missed out on the opportunity to appreciate many of the factors that have made them so successful as a species.

The first concerted attempts to understand krill started in the 1920s. The impetus for these studies, rather than being curiosity about the biology of these fascinating creatures, was an attempt to understand their major predators—the great whales. There were many natural history observations of whales made from whaling ships, but most of the information the whaling industry gathered came from observations of dead whales made on ships or at whaling stations. Whale stomachs opened on the slipways of the whaling stations poured out metric tons of krill. The large baleen whales—the fins, blues, and humpbacks—were almost entirely dependent on Antarctic krill for their food. Whale populations were known to fluctuate, and their abundance could change dramatically from year to year. It was assumed that this reflected the cycles of abundance of their prey. It was logical that to understand the biology of these whales—and to increase the profitability of the whaling industry—knowledge of their prey was essential. So began the Discovery Investigations, funded by levies on the whaling industry, which focused on understanding Antarctic krill but which also shed light on a host of other Southern Ocean organisms and processes. For most krill biologists, the highlight of the thirty-seven volumes of research that were published between 1929 and 1980 was *The Natural History and Geography of the Antarctic Krill* (Euphausia superba Dana) by James Marr. This 463-page monograph described most aspects of the biology of krill derived from a series of sampling voyages between 1926 and 1940. This weighty volume included analyses of nearly 12,500 samples and included measurements of krill and descriptions of their probable life history derived from these measurements. Marr also included lyrical descriptions of living krill and some well-crafted commentary on the issues of conducting research on krill. I have liberally borrowed his lucid words throughout this book.

James Marr was a highly unusual scientist. As a Boy Scout he had sailed with Ernest Shackleton (1874–1922) on the ill-fated last expedition where the great polar explorer died of a heart attack. He studied classics at Aberdeen University (he was a Scot) but later switched to zoology. His love of the classics is legendary; according to some reports, he would retire after a long day sampling in Antarctic waters on the *RRS Discovery* by lying on his bunk and reading Latin poetry. Later in life he would lead a secret wartime British expedition to

Antarctica. His unusual background led to an ability to write compelling prose, and his huge volume is incredibly readable. For many modern scientists, the work seems overly descriptive, but I was entranced when I first encountered his writing. He could assemble a vast array of facts and provide a synthesis based on measurements and on his own observations of krill and their environment.

Since the early twentieth century our knowledge of krill biology has slowly accumulated, but we are still hampered by their remote and difficult habitat, the limited techniques available, and always by their fickle biology. To understand how krill cope with their watery environment we need some basic information regarding how fast they grow, how they reproduce, how long it takes for them to reach maturity, how long they live, and the nature of their seasonal cycle. Underlying these details is the great unknown—krill behavior. How krill behave can affect all the measurements we make and can undermine many of our most basic assumptions. Krill behavior is perhaps the most critically understudied aspect of krill biology, partly because it is so difficult to assess and so is easily ignored.

Nowadays most krill research is conducted off well-equipped research icebreakers bristling with electronic sensors that provide the best chance of locating and catching krill. This was not always the case. On my first Antarctic expedition, we had to use a ship of convenience (we had to improvise because our research vessel had struck some rocks and sank). Because vessels equipped to deal with ice are rare in Australia, we had to borrow them from the Northern Hemisphere. Thus we used a chartered Canadian ice-strengthened freighter, and we equipped the vessel as best we could for krill research. We modified a battered and dented refrigerated shipping container, chained it to the hatch covers, and used it as a makeshift laboratory. A handheld immersion pump lowered into the heaving ocean provided us with our seawater supply, and towers of buckets stacked along the walls of the container functioned as our aquarium. To catch krill, we had to wait for night when we guessed they would be closest to the surface (the ship lacked the echosounders that could have pinpointed krill aggregations) then lowered a small hoop net off the ship's stern as it drifted in the blowing snow among the icebergs. Hauling the net required press-ganging any idle ship's passenger, and I was joined in this task by traveling journalists, government bureaucrats, army boat drivers, and the ship's Newfoundland crew all eager to see what would emerge from the depths. Net retrieval involved hauling hundreds of meters of cold, wet rope onto the slippery deck by hand. To my surprise, and to the delight of my motley band of assistants, we managed to get several buckets of live krill into our ramshackle "laboratory," and even back to Australia, using this labor-intensive method.

Most krill research programs are more professional these days. There are several large (roughly a hundred meters [110 yards] long) icebreaking research vessels in use by national Antarctic programs. They have the scientific echosounders necessary to detect swarms of krill and the winches and hydraulics required to deploy and retrieve large nets and other equipment. Even today, many studies merely use a net to sample for krill (and other creatures) at a series of fixed positions and preserve the collected animals for later study, just like scientists have done for a hundred years. More ambitious studies require the collection of living krill, and several ships are now set up with dedicated onboard aquaria/laboratories where krill can be kept alive and can be experimented on—and then can be brought back to land-based laboratories in temperate regions for further study. Many modern research voyages combine the two approaches and illustrate a change in thinking between the historical approach to marine biology and a more advanced attitude that acknowledges the need to understand more than just the demographics of the krill population.

Sampling krill requires a plan. Most surveys follow a rigorously designed series of transect lines and sampling stations, and the equipment is deployed to set protocols. Our eyes in the ocean are the scientific echosounders. These look down below the ship and tell us if there are any aggregations of animal life in the depths. Scientific echosounders have been used regularly since the 1970s and have become the standard tool for scientists studying the distribution of the larger animals of the pelagic zone, including krill. Echosounders work by sending pulses of sound down into the water, then recording the echoes reflected from objects in the water. By using different frequencies of sound, we can separate the signal that comes from krill

A printout from a scientific echosounder illustrating the two-dimensional structure present in a swarm of Antarctic krill. Darker areas are where the krill are most dense. White indicates an absence of krill. The sea surface is at the top of the picture and the seafloor at the bottom. The ship was moving across this swarm from right to left.

from those generated by other groups of animals. With experience, it is possible to inspect the echosounder screen and determine which returns are from krill and which are from other organisms, although it is prudent to occasionally test these assumptions by turning the ship around and towing a net through the red blob we detected on the echosounder. These exercises are excuses for some competitive activity onboard the ship; we all have different expectations of what will emerge from the net, and many wagers are won or lost when the catch comprises half a metric ton of jellyfish instead of the much-anticipated krill.

The real value of the echosounder is that it can quantify the abundance of krill under the ship and provide a continuous trace of how much krill is along the track line. Echosounders have revealed the spatial complexity of krill aggregations and their huge size. This continuous sampling of the water generates vast amounts of data, which can be stitched together to provide a three-dimensional map of krill distribution and abundance in an area. Understanding the distribution and abundance of krill is key to unlocking the secrets of the ecosystem—why krill predators go to feed in certain areas, how krill are linked in to chemical and physical features in the ocean, how krill utilize the food available in their habitat, and why krill are present in some areas and not in others. Estimating how much krill is in an area is also a prerequisite to providing advice on managing the fishery.

A one-off measurement of the abundance of krill is of limited use. To detect change between seasons or years we must measure abundance multiple times in the same place. This requires considerable commitment of the resources to conduct numerous surveys in an area year after year. We also must have confidence that we can measure the distribution and abundance of krill accurately and repeatedly. This is a big ask. One of the defining characteristics of krill aggregations is that they change rapidly. This change can be in depth, shape, or density. We really don't understand why this happens but it does mean that steaming over a krill aggregation once will get one picture of krill distribution and abundance, but steaming back over the same track shortly afterward will generate an entirely different picture. We need to be able to capture this variability and take account of it when we compare the results of different surveys. Again, this is very difficult.

This uncertainty in our estimates of abundance does not prevent us from determining when a major change has occurred. At South Georgia, near the northern limits of krill distribution, there are years when krill are very difficult to find. In these "bad krill years" scientific surveys show very low levels of krill density and abundance, fishing vessels move to other areas, and krill predators from colonies on the islands starve. The fact that all these independent measures of krill availability show the same pattern gives us some confidence that our scientific surveys can detect massive changes in krill availability. It is the subtler changes that we are not yet able to pick up with any degree of confidence. Uncertainty like this is not unusual in complex systems. We are all familiar with the errors associated with the results from political polls, and statisticians have developed a suite of techniques that allow us to determine the level of confidence we can have in our estimates. It is not enough to have just an abundance estimate from a survey; we also need to know the error associated with that estimate. This is often difficult to pin down accurately.

Echosounders have no idea about krill—we have to educate them. Krill hit by a sound beam give off a characteristic echo. This echo changes depending on the water temperature and on the swimming angle of the krill, their physiological state, their size, and how many other krill are around them. We determine each of these characteristics through onboard measurements and from experimental studies. This information is then fed back into the algorithms that are used to translate the electronic information received by the echosounder into measurements of krill density (the number of krill in a given volume of water). Obviously, we do not have the complete set of parameters, newly collected, to calibrate each echo. We use averages, or ranges of values, to translate the echoes into krill density. Considerable research goes into improving this translation process, so the confidence in our density estimates is improving with time.

That said, regardless of how much we improve our processes it is still an exercise in estimation rather than direct measurement. Terrestrial biologists can lay out a transect across a pasture and directly count each individual beetle they see along its length. The errors in their estimates of density, then, are a product of how they extrapolate from their measured one-dimensional transect to a two-dimensional area. In the ocean, we cannot count krill directly so the errors involved in our density estimates include those inherent in the measurement device—the echosounder or net—as well as those that arise from extrapolating from a two-dimensional transect line to a three-dimensional volume. If we include time as another dimension, then life becomes still more complicated. This is all to say that we operate in a sea of uncertainty and that the best we can hope for is estimates that *approach* the real value. Importantly, we want to be sure that estimates derived from one time and place are comparable to those collected from others.

Given these reservations about accurately measuring krill distribution and abundance how can we have confidence in the assertion that krill may be the most abundant multicelled animal species on Earth?

We know that krill occur over a vast area, and we can use records from decades of net surveys to specify the minimum boundaries of that area. We can use density estimates from the nets and the acoustics to provide a range of plausible densities that might be found within the krill habitat. This process gives us minimum and maximum estimates of overall abundance—but given the tendency of both estimation techniques to underestimate krill density the result is a conservative estimate. This process suggests that the total tonnage of krill in the Southern Ocean is around 379 million metric tons.

There are other, more indirect, ways of estimating how many krill there are or how many krill there could possibly be. It is much easier to count seals and penguins than it is to count krill. We know that most of these species in the Southern Ocean depend on krill so we can get an idea of how much they could possibly consume. We can also estimate the consumption of krill by all the other krill predators—the whales, flighted seabirds, fish, and squid—and we can add these together. If our assumptions on the numbers of these animals and their rate of krill consumption are correct, then there would have to be at around four hundred million metric tons of krill in the ocean. Of course, predators consume only a proportion of the annual production of the krill population, so estimating krill abundance by predator consumption gives another conservative estimate of krill biomass. Finally, we can look at how much energy there is in the Southern Ocean and estimate the proportion of that energy that is available to krill. Algae are the source of the energy available to herbivores like krill. We can calculate the algal production in the Southern Ocean and then examine the relative consumption by all the different herbivores. Such calculations indicate how much krill the Southern Ocean could support, which could be as high as four billion metric tons. Given the assumptions and margins of error around this figure, it is very much an upper limit.

When all these numbers are taken together it becomes obvious that krill are enormously abundant. The biomass of any species depends on the individual weight, the density in which the individuals live, and the area over which they are distributed. Because krill are relatively large for marine animals, are densely packed, and occur over a huge area, they naturally have a high total biomass. There are other marine animals that are hugely abundant and widespread, but each one is only a small fraction of a krill's weight so they cannot compete in the biomass sweepstakes. On land, there are many abundant large animals, but the land is only thirty percent of the planet's surface area so they are at a disadvantage. We do have a good idea of the abundance of many land animals, and in terms of overall biomass cows (650 million metric tons) are probably now the dominant land animal, followed by humans (350 million metric tons). Our various estimates for krill biomass are spread across this range. Most krill experts would agree that, given what we know from all sources of information, around 500 million metric tons is a realistic biomass estimate for krill, which means that they are among the most abundant species on the planet. But I'm not happy with second place, so I like to state that there is little conclusive evidence to suggest that they are *not* the world's most abundant multicelled animal.

~

Electronics and computers have revolutionized marine research over the last thirty years. Echosounders are now the gold standard for assessing krill distribution, but there are problems with looking at historical acoustic data. Acoustic surveys of krill conducted in the 1970s and early 1980s used echosounder equipment and analysis software that differed greatly from those in use today. It is thus impossible to convert and quantitatively compare those early data with the data we currently gather. Our acoustic data from the 1980s reside in boxes of computer printouts, whereas the data from our more recent voyages leave the ship on a hard drive or a memory stick. There is, however, one abundance estimation technique that has changed little in the last 150 years—nets.

Much of the biological information we have gathered about the oceans has been obtained by towing nets through the water, and the process has changed little over the years. Dipping a net in the water to catch krill seems like a straightforward activity, but it is not. The nets produce samples of animals that are then preserved in formalin and examined later in the confines of a nice, warm, stable laboratory.

Krill research in the 1920s—only slightly more primitive than current practices. From the left: the Australian polar explorer Douglas Mawson; Dr. W. Ingram, medical doctor and bacteriologist; and the great krill biologist James Marr. (AAD photo)

So for most of last century most of what we knew about krill came from dead animals. Theoretically, nets avoid many of the complicated problems that emerge when the electronic signals from echosounders are converted into meaningful biological information so they still have many supporters. Nets produce actual samples of recently living animals that can be identified and measured. But the type of net used and the way it is used causes problems.

Nets were originally used to sample the ocean when we knew little about how marine animals were distributed. If an open-ocean species is small, is of limited swimming ability, and is distributed randomly, then a net would be a great sampling tool. We now know that few marine organisms meet all these criteria. We do know that krill are large, fast swimming, and nonrandomly distributed. Krill can evade slow-moving nets, and nets can easily miss areas where krill are most abundant, so average densities calculated from nets are always underestimates of the real value. We can see from echosounder records that

A small-mouth, fine-meshed plankton net, which is not well suited to catching fast-swimming krill. (Photo by author)

krill swarms have great structure—there are patches of very high density separated by empty water, and a range of intermediate densities as well, all showing variation in three dimensions as well as in time. In some areas, over ninety percent of the biomass of krill is present in these tight aggregations so getting the density estimates correct is critically important.

Estimating density of krill from samples collected by a net is deceptively simple. Scientific nets are equipped with flow meters to measure the volume of water that has been sampled by the net. We can count the number of krill retained in the cod-end, and then a simple sum produces a density measurement—the number of animals per unit volume. The complications arise because nets differ in their mouth opening, mesh size, and length, and they can be towed in different ways, at different speeds, and to various depths. Some researchers tow the net vertically, some horizontally and some obliquely. Because of the patchy nature of krill distribution, these subtleties can greatly affect the number of krill caught and hence the estimate of density. These problems are less serious when one type of net or technique is used consistently in a given study, but complications arise when one is comparing results between studies that have used different nets in different ways. In one recent study looking at long-term change in krill populations the dataset used results from over twenty different net types and all possible ways of towing the net. Disentangling signals from such information is fraught with difficulties. Using a towed net to understand ocean ecosystems has been compared to trying to understand the Amazon rainforests by flying over the treetops in an airplane, waving a butterfly net, while blindfolded.

The main problem we have with measuring krill density is that we have very few ways in which we can ground-truth our estimates. Currently we can compare our estimates of density from echosounders with our estimates from nets, but we inevitably get different results; echosounders provide estimates of density that are higher than those produced by nets. Estimates of krill density made by scuba divers are usually ten to a hundred times higher than the values from echosounders or nets so there are major discrepancies that need to be sorted out. Determining the real density of krill in the ocean is a significant problem that affects our ability to understand krill, their ecosystem interactions, and the effects of the fishery. Some clever research is required to solve this problem. We need an independent and more precise estimate of density that we can compare to the values we get from nets and echosounders.

Back when I was working in the Bay of Fundy, I had the distinct advantage of being able to see the distribution of the krill at the surface and I could perceive the great differences in density that occurred in a very small area. Back then I had a crack at seeing if I could compare density estimates from different sampling devices. I had no echosounder (this was a low budget operation), but I did tow a net through an area that was quite obviously teeming with krill.

The krill were present in many dense swarms separated by areas where krill were scarce or absent. The net produced woeful results— the average density was estimated to be fewer than six krill per cubic meter (1.3 cubic yards), and most tows yielded no krill at all despite the obvious presence of dense swarms throughout the area. I then estimated the density of krill in swarms from some photographs I had taken by measuring the distance between the individual krill in the photographs. Densities measured this way ranged from 9,000 to an astonishing 770,000 krill per cubic meter (1.3 cubic yards).

Finally, I had a rather unusual, low-budget sampling device, which consisted of a dry-cleaning bag sheathed onto a ring at the end of a pole. Some unkind people likened this device to a condom, but it was actually derived from observations of feeding baleen whales. The idea was that you lower this device into the water and slide it over the swarm, thereby capturing the water and the constituent krill. Because I knew the volume of the bag, and I could count the number of captured krill, I had a direct estimate of density. These estimates ranged from four thousand to forty-one thousand animals per cubic meter, much higher density values than any species of krill measured using conventional techniques, but lower than that estimated from photographs. I had verified that krill live in very high densities in swarms and that towing a net behind a boat was probably the worst way of estimating krill density. These conclusions have tainted my approach to sampling the ocean ever since.

The ability of krill to evade nets is not news. In 1934 one of the Discovery-era scientists, Mackintosh, wrote: "I have been able from the ship's side to watch a net being towed a few feet below the surface and passing through a shoal of this species [Antarctic krill]. The Euphausians could clearly be seen to leap backwards out of the way." These observations led the Discovery scientists to try using boom nets towed from the front side of the ship, thus avoiding the disturbed water at the stern.

It is not possible to sample large volumes of the ocean using dry-cleaning bags, but more sophisticated technology is on its way and may assist in years ahead. These will include improved three-dimensional echosounder technology and advanced video and digital camera systems, and hopefully more concentration on studying krill in their ocean home. There are now even suggestions that the hoary old problem of krill distribution and abundance could be tackled by using advanced molecular techniques to detect the DNA signature that krill leave in the seawater. Could a pipette replace nets and echosounders in the future?

For now, the bottom line when it comes to estimating krill, density, distribution, and abundance is that both of our major sampling tools (echosounders and nets) have a significant number of drawbacks. This does not mean that we are unable to assess krill distribution and abundance, it just means that we must be well aware of our limitations and very cautious in our conclusions. I am frequently asked whether krill populations are changing, and it is frustrating to admit that I really don't have an accurate answer. Because krill occur over such a huge area we can never census the population sufficiently to be confident that changes we see are a result of changes in abundance rather than changes in distribution. An apparent change in krill abundance in an area may just result from the krill moving to a different area. We also suffer because of the rapid rate of technological change in our most reliable density estimation technique, echosounders, and because of the problems associated with using nets to understand krill distribution and abundance. Consequently, reports of major declines in krill abundance must be scrutinized very carefully because the results are so dependent on the technique used. I am yet to be convinced that we have the data to unequivocally make any statements about long-term trends in the krill population. Hopefully technology will come to the rescue.

Modern electronics are allowing us to see the world of krill as never before. Echosounders have given us a picture of what krill swarms look like, albeit in two dimensions, and have allowed us to more accurately count krill. More recently, underwater video has been giving us a window into the behavior of krill in the dark heart of the ocean. For the most part, krill (of all species) live in the ocean's interior and are rarely seen. For the last fifty years, marine ecologists have focused more on quantitative methods and models at the expense of visual observations. Despite this blind spot, many of the recent eye-opening discoveries about krill have come from quasi-accidental visual observations.

Diving in the open ocean, and particularly in the Antarctic region, is fraught with dangers, which explains the paucity of krill observations by scuba divers. Organizing a diving expedition to the Antarctic is also costly and usually involves taking a decompression chamber and a large team that includes a trained diving medical specialist. That is, of course, on top of having a large ice-capable ship from which to mount the expedition. Not surprisingly, there have been very few such expeditions, but these have produced a disproportionate wealth of observations, many of which have changed the way in which we look at krill. Divers have observed and documented the close relationship between krill and the complicated topography on the underside of the ice in winter, photographing the characteristic upside-down feeding behavior as the krill scrape the algae off the bottom of the ice. Their observations, revealing the three-dimensional cloudlike structure of krill schools, have allowed us to interpret the results we see on the screens of our echosounders. Humans are visual animals, and seeing krill individually and in schools provides us with the context we need to put our other, more remote studies into context.

Using a submersible would be an ideal way to observe krill behavior, but it is a more complicated endeavor still. There have been only a few observations from manned submersibles in the Southern Ocean, but these have reported seeing krill. There are, however, observations from submersibles in other parts of the world that have provided new insights into various aspects of krill behavior. I was lucky enough to make two dives on the Canadian submersible *Pandora II* off the Atlantic Coast of Canada. We were looking for North Atlantic krill to examine whether we could find them in densities like those we were encountering on the surface in the Bay of Fundy. At first, we were disappointed because the waters near the surface were devoid of krill. As the submersible slowly sank into the murky depths we filmed a range of small ocean creatures but few animals that were unequivocally krill. Then something strange happened. As we approached the silty seafloor at a depth of about three hundred meters (328 yards), we encountered dense aggregations of krill, just in the few meters above the seafloor.

Finding krill at three hundred meters was a surprise. The picture that scientists had built up over the years sampling these krill from oceanographic research vessels was that they were only found in the top two hundred meters (219 yards) of the water. But how did we know this? From using nets, which were so fragile that no one would dare to use them to sample the water near to the seafloor and by using echosounders that had a range of just two hundred meters from the surface. The tools we possessed were only capable of finding krill when they were close to the surface so our concept of where krill lived was, not surprisingly, biased toward the surface layer. I learned a valuable lesson from this experience—our knowledge of life in the ocean is only as good as the tools we use to explore that part of the ocean where we cannot see. James Marr, in his book on Antarctic krill in 1962 knew this, writing, "for too long we have been putting too much trust in what *seems* to be revealed by apparatus and not enough on what we can *actually see*." Unfortunately, this observation is still true today.

Considerable Antarctic research is now conducted using remote cameras or underwater vehicles, where the main aim is not to investigate krill behavior. To these researchers, the presence of large quantities of krill frustratingly obscures their view of the true object of their research. In one such study, some of my colleagues were attaching underwater video cameras to fishing gear designed to catch bottom-living fish on the Antarctic continental shelf. They were examining whether long-lines would have a detrimental effect on the ecosystems of the seabed, including the sponges, deep sea corals, and other sessile organisms that were delicately attached to the bottom. Every time they sent the fishing gear to the seafloor, they were frustrated by the images on their video camera being obscured by abundant krill. Amazingly, the krill were filmed at six hundred meters (656 yards), far deeper than most of us expected them to be found. But the real treasure was to be found when we examined the video images in detail—there were the first ever clear images of krill engaging in sex. Two males were seen chasing a female, and eventually one of them achieved his aim. The sequence of movements we described from the footage were delicately referred to as chase, probe, and embrace. We worked with a talented animation artist who produced the first pornographic krill film, which was a huge hit in Antarctic circles. The animations found their way into the scientific literature too, and one of the proudest moments of my scientific career was the resulting publication, with the title "Ocean Bottom Krill Sex!"

Underwater video and photography are now being used much more frequently on oceanographic sampling gear, and more images of krill in the ocean are emerging. Examination of kilometers of video footage in search of krill has led to one of the emerging paradigms in krill biology—that the krill population may live at great depths in Antarctic waters—even down to four and a half kilometers (2.8 miles) from the surface. Of course, the real question is whether krill have always been there or whether this is a new behavior that has resulted from some oceanic change or human disturbance. Because of the lack of historical images, we will probably never know.

With the rise of Antarctic adventure tourism more people are snorkeling and diving in Antarctic waters than ever before. Most of these adventurers go south in search of images of seals and penguins, but some accidentally encounter krill, and those images can find their way onto video sites. One such sequence filmed on a snorkeling expedition in the Antarctic follows an individual, heavily pregnant, female krill as she swims through fronds of kelp in coastal waters off the Antarctic Peninsula (https://www.youtube.com/watch?v=OPMQaP-YjɪY). I have always been curious about how individual krill spend their time in the wild, and this video is a partial answer. The krill swims in a relaxed fashion through the water, stopping to investigate some small object, and then goes about her business. She swims purposefully, spiraling up and then down, slowly sinking down while using her swimming legs as parachutes, looping-the-loop, hovering, tail flipping when she bumps into some seaweed, and all the time being apparently oblivious to her photographer. If I want to convince a doubter that krill are curious, active, behaving animals, then I show them this video clip, and they are usually converted.

Swimming with krill is a rare and treasured event that is available to few. The development of new technologies that will allow us to better study krill in their natural home should allow us to overcome many of the problems that limit our understanding of the biology and ecology of krill. Many of the remaining problems will be conceptual in nature—how we, as scientists, view krill.

In science, there is a natural divide between "lumpers," who find ways to combine biological groups together and "splitters," who view each group individually. Biological oceanographers have traditionally been lumpers, and most group krill together with almost all living animals in the sea under a single term, *zooplankton*. But what are zooplankton, besides tiny twitching things you can see in a drop of seawater under a microscope? Dictionaries define zooplankton as animals of small size that float and drift in the water layers—basically, small aquatic animals that do not swim well. This description applies to most animal life on the planet but ironically not to adult krill, which are not small and can swim quite well.

I dislike the term *zooplankton* because it lumps together most forms of life in the ocean under a single grouping term without regard to their method of feeding, their behavior, their body type, or their evolutionary status. In terrestrial systems, which we understand far better than marine systems, there is no such grouping term. Imagine if we lumped together all the bees, ants, slugs, centipedes, beetles, and worms under one term?

Perhaps what is most insidious about a term such as *zooplankton* is that it comes with considerable baggage. When we label krill as zooplankton they are immediately assumed to be tiny, feeble swimmers—they are consigned to the soup of tiny animals that slosh around unseen in the ocean. This perspective makes it fiendishly difficult to garner enthusiasm for the conservation of krill and other similar animals of the open ocean. There is an inverse relationship between the degree of concern felt by the public and the size of the organism. The conservation of whales, elephants, rhinos, tuna, lobsters, octopuses, or turtles attracts considerable support. It has been much more difficult to achieve support for campaigns to save bees, protect butterflies, or conserve krill. The problem for krill is that, though they are genuinely smaller than the "charismatic megafauna," most scientists and commentators use a term that implies that they are smaller still and almost inanimate. If we fail to face up to the true animal nature of krill in our studies, we will continue to make mistakes and to underestimate the abilities of this marvelous organism.

Space is often called the last frontier, but in reality, space is abstract—we can't see it, we can't go there, and there's not a lot we can do with it. The oceans, on the other hand, begin at our coastlines; to many of us these are our actual frontiers, yet what goes on in the vast oceanic volume is almost a complete mystery. We use the oceans—we travel over them, we swim in them, and we extract food and minerals from them, but they are remarkably difficult to understand. We are terrestrial creatures; we need air to breathe and ground to stand on. In the ocean, we lose the ability to breathe, to see properly, to move freely—and this is just in the tiny section of the ocean that we can access.

Mostly the ocean is as cold, dark, and lethal as space. It amazes me that we spend only a small fraction of the amount that is invested in the space industry in exploring the much more accessible mysteries of the ocean. The oceans are critical to life on planet Earth. They produce much of the oxygen we breathe; they absorb many of the poisons we pump into the atmosphere, rivers, and the seas; and they are the source of the rain that allows plants to grow on the thirty percent of the planet that is above water, thus making our terrestrial lifestyle possible.

We have treated the oceans like a bottomless larder and as a planetary-scale garbage dump where our waste products end up—out of sight and out of mind. The volume of the oceans is beyond our ability to grasp, yet we have managed to have an impact in even the most inaccessible corners of the watery domain. To detect life on distant planets we have developed satellites, telescopes, planetary rovers, and rockets, but to explore life in the oceans we are still reliant on blindly towing nets behind ships, just like the earliest ocean scientists did 150 years ago. Unfortunately, many of our concepts of how animals like krill live are almost as archaic as the instruments we still use to study them. Future progress will require a new attitude as well as new technology.

*Chapter Four*

# Bringing Krill to Life

*E. superba*, far from remaining a passive drifter, has on
the contrary become a creature of great agility, powers of
locomotion, purposeful intent, and not a little awareness.
—James Marr, *The Natural History and Geography of the
Antarctic Krill*

The beauty and vital nature of krill obviously captured the imagination of the early krill scientists working in the 1920s and '30s. Despite the vibrant descriptions of schooling krill that bring their writings to life, their science was almost entirely focused on routine measurements of preserved animals. There was one exception. A paper published in 1967 described an experiment on live krill carried out in the 1920s. The author seemed almost apologetic for wasting the reader's time: "The observations described in this paper were made only as a casual experiment to occupy a little spare time." Obviously, understanding the biology of living krill was not deemed that important back then. It wasn't until the 1960s, when that paper finally saw the light of day, that scientists felt the need to devote a little more time to the study of living krill. In the 1970s and '80s the spotlight slowly turned to Antarctic krill, largely because of the growing fishery (see chapters six and seven). The largest ever international marine biological research program, Biological Investigations of Marine Antarctic Systems and Stocks (BIOMASS) commenced. This program focused on the role of krill in the Southern Ocean ecosystem. BIOMASS invigorated studies on living krill, and the modern era of krill biology began.

Several groups of scientists began experimenting with live krill in the 1960s, first in primitive laboratories onboard ships and later in dedicated laboratories on the Antarctic continent and at South Georgia. These early studies established the basic conditions necessary to keep krill alive and happy. In the 1980s the facilities available on research vessels were becoming more sophisticated so Antarctic krill could be studied onboard. Krill could also be brought back alive to laboratories on the Antarctic continent and to scientific institutes in Europe, South Africa, and even tropical Australia—a considerable logistical feat. Some krill were even flown back to Germany as a scientist's hand-luggage allowance on a commercial airline flight. In these laboratories, teams of scientists developed innovative studies into all aspects of krill biology, physiology, biochemistry, and behavior. Some of these investigations were based on techniques that had been developed for studying more accessible species of krill in the Northern Hemisphere. Species that lived in the coastal waters of the Atlantic and Pacific Oceans could be caught and rapidly returned to the shore-based laboratories. What would have taken months in the Antarctic could be achieved in hours in more temperate waters, and the lessons learned could be generalized to other krill species. We began to learn that krill, of all species, rather than being too delicate for laboratory study, were in fact quite robust—once we had mastered the techniques for their transport and maintenance.

These developments nurtured a group of young scientists who were enthusiastic about krill research. Some approaches to working with living krill were more successful than others. I have happy memories of my early attempts to bring live North Atlantic krill back from the Bay of Fundy to the laboratory in Halifax. This involved a hair-raising dash across Nova Scotia in a van with several barrels full of seawater and hundreds of highly confused krill. Back at the laboratory, the krill

Most people's introduction to krill: dead, limp, and smelling of formalin.
(Photo by author)

were not very compliant experimental subjects, which I attributed to car sickness rather than to my rather shaky experimental methods. These early experimental studies allowed researchers to develop experience in krill-husbandry methods that resulted in far more relaxed krill—and krill researchers.

Many of the key breakthroughs in krill biology came about through studies of living krill in aquaria, and developing techniques to capture and study living krill has been a focus of much research over the last forty years. Krill are not laboratory rats, and studying them alive has required ingenuity and creative problem solving. Today, we use especially gentle collection methods to provide krill for aquarium studies. This requires towing a large net very slowly, often on an aggregation that we have detected using the echosounder. The captured krill are brought gently aboard, hopefully alive, and are available for study. This sounds simple but in fact it took years of practice before we had mastered the technique. Krill brought aboard this gentle way are amazingly lively, frothing up the bucket of seawater they are emptied into. A too-vigorous trawl and we end up with a catch of limp and listless krill, which are useless for observations or experiments.

Getting krill from the Antarctic to a laboratory in more temperate climes is a logistical nightmare. Krill are most comfortable in seawater temperatures of $-1.5$ to $5°C$ ($-34.7°F$ to $41°F$). This is not a problem in Antarctic waters because the holding tanks can be irrigated using seawater directly from the cold ocean. On the northward journey, however, the seawater warms, and within a few days it can reach $6°C$ ($42.8°F$), which is too hot for krill so special provisions must be made for chilling large quantities of seawater for the return trip.

Once safely in the temperate latitudes, the krill need a new home. They need a facility that can keep them cold, well fed, and with appropriate lighting. It took us years, and not a few dollars, to develop a system that allowed us to keep krill alive at our institute in Australia. Initially we conducted our research in tanks and buckets kept in cold rooms, chilled to $0°C$ ($32°F$). This meant that to carry out experiments and maintenance we had to wear full Antarctic clothing despite it often being warm and sunny outside. Working at low temperature worked well for krill but made for a dangerous work environment for

The krill research aquarium at the Antarctic Division's headquarters in Tasmania. Many thousand krill are kept in large holding tanks, and smaller containers are used for experimental treatments. Schooling krill can be seen in the tank in the foreground. (Photo by Rob King, AAD)

scientists and technicians. The solution was to construct a dedicated facility with chilled seawater but ambient air temperature. Not surprisingly, scientists were more keen to work with our krill once we had civilized conditions for both researchers and crustaceans.

But cool water temperature is not the sole requirement for a contented krill: they are very sensitive to water quality. Because krill feed ravenously, they produce a lot of liquid and solid waste, which quickly contaminates their new home—we needed to build in elaborate mechanisms for water purification. Krill are also very sensitive to external contaminants. Our aquarium was established at our institute in Tasmania, a relatively undeveloped island, off an island continent at the far end of the world. We naively imagined that if we collected local seawater it would be sufficiently pure for our newly collected krill. To our surprise, the krill reacted badly to the level of contaminants in water that we imagined was pristine and unpolluted. Today, the Antarctic Division trucks in seawater from a distant site, and all the seawater in the aquarium is given several levels of filtering and treatment. Once we had achieved good water quality, we had happy krill and could study other aspects of their biology.

Our research aquarium was, for a while, the only place where live krill could be observed and studied outside the Antarctic. This fact did not escape the attention of large public aquaria in Japan, and we assisted these institutes by providing krill and expertise so that the public could, for the first time, see and learn about the wonders of krill. But the flow of information was not all one way. The Port of Nagoya Public Aquarium used its considerable resources to refine the procedures for keeping krill in captivity—improved diets, lighting, and water purification systems. These advances allowed their krill to grow and even reproduce—aspects of their biology that we had struggled with. They shared this information with us to improve our research facilities. Suddenly, our tame krill were happy to school in their tanks, to grow and reproduce, and even to produce several generations of offspring, which in turn produced their own progeny. We now understand that krill, when given certain basic requirements, are quite robust experimental animals. They may not quite be laboratory rats, but they are as adaptable as we might expect given what we know about their biological success.

But what can be learned from living krill that could not be as easily gleaned from measurements of preserved animals? Obviously, there is an almost infinite number of interesting studies that we could carry out with a thriving population of krill in a laboratory, but we need some key information to appreciate how krill cope with their environment and how they carry out their day-to-day life. Some of this information is also critical if we are to manage the krill fishery sustainably. Understanding how old a krill is and the rate at which individuals and populations grow is one of these key areas of knowledge. Since the first investigations into Antarctic krill commenced in the early twentieth century, growth and age of krill had always been determined from length measurements of individual, formalin-preserved samples caught with nets deployed from research vessels. The underlying assumption has always been that bigger krill are older krill, so age could be determined by measuring the length of krill. This was, and still is, standard practice for many crustaceans. Any large sample of krill collected from the same population was observed to have several size clusters. These, it was assumed, were age classes; one cluster was produced every year, and the number of clusters indicated the maximum age of krill in a given population. The collective wisdom after thirty years of measuring the sizes of preserved krill was that there were two size groups in the adult population; thus it was assumed that krill probably grew quite rapidly when young but matured and died when they were about two years old. But no one had ever witnessed this process.

That first recorded experiment on living krill back in the 1920s had documented how krill grow. The body of krill, like that of all crustaceans, is covered by a shell, or exoskeleton, which gives the body its shape and provides a hard substrate for the muscles to attach to. The shell is both rigid and external; thus krill face a problem when growing because the entire body is constrained by the size of the shell. The solution is to cast off the old shell and replace it with a new one that has been growing under its predecessor—a process called molting. This is what all crustaceans do, but in some of those first aquarium studies in the 1980s it was discovered that in krill there was a novel twist that changed everything we thought we knew about krill age. Adult Antarctic krill, when kept at 0°C (32°F), molt monthly throughout their life. To grow, they produce a slightly larger new exoskeleton, which, after molting, they grow into by pumping water into their bodies and stretching the new shell before it has hardened. The confounding twist that krill displayed was that, if food is scarce, krill will produce a new shell that is smaller than the old one, and they shrink rather than grow during molting.

The first study that showed shrinkage was trying to answer another question altogether, concerning how krill survive the long, dark, food-scarce Antarctic winter. Could krill endure if there was little food around? The answer was yes, they could survive for a spectacular 211 days without food. During this extended fast, the starved krill lost half their body weight, using up their fat reserves and some of their body protein. But what was truly eye opening was that they continued to molt regularly, even in the absence of food, and instead of growing in length, they got shorter and smaller with every molt. Not only did they shrink, both male and female krill began to lose all the external evidence that they were, in fact, mature males and females, and after several months they were indistinguishable from juveniles.

This was more than just a neat trick to get krill through winter. What this meant was that just because a krill appeared to be small and immature didn't mean that it was a young krill. When it became apparent that small krill could, in fact, be shrunken old krill, scientists began looking at the size structure of the population more carefully. Sure enough, the adult population appeared to consist of a much larger number of year classes, except they were merged together as might be expected if the individuals were growing and shrinking throughout their lives. It seemed likely that, in the wild, krill could live for at least five years—two and a half times longer than originally thought. More startlingly, in the laboratory, they could live for much longer.

The most straightforward way to find out how fast a krill grows and how long it can live is to measure the size of captive krill as they grow. This approach became possible once krill were being successfully maintained in the laboratory. Japanese researcher Tom Ikeda, working in Australia, was a genius for keeping krill alive. It was he who, in the early 1980s, had shown that krill could survive over two hundred days of starvation. Tom kept his krill in individual jars in a cold room. He was meticulous, checking each krill daily, changing their water and removing waste material and discarded molts. The molts proved extremely useful because, by measuring their size, he could work out how big the krill was before it molted. He accumulated months of these measurements and was able to track individual growth and shrinkage over a year without overly stressing the krill. Tom nurtured each of his krill and even named them all. As his krill got older they became more valuable because they had already been alive in the laboratory for many years, exhibiting astounding longevity. His prize specimen, Alan, had been swimming around in his two-liter jar in the cold room for nine years. Alan had probably been at least two years old when captured, which meant that he was technically over five times the accepted age for any known krill at the time. Unfortunately, Alan met an untimely death when he accidentally escaped during a water change and disappeared down the drain, so we will never know the ripe old age he could have attained. His legacy is that his record-breaking longevity changed forever how we thought of krill. No longer did krill grow fast and die young; we now appreciated that they took two years to mature and then could live long and fulfilling lives, drainpipe accidents notwithstanding.

The implications of the new-found longevity of krill were profound; the vast krill population in the Southern Ocean was the product of many years' production, not just two. The krill population was much less productive and robust than we had imagined. These findings appeared at the time when the fishery was beginning to expand and resulted in a far more precautionary approach to management than would have been the case without Alan's sacrifice (see chapter seven). No one since has had the patience to keep individual krill alive for over nine years to see how old they might be able to grow and to break Alan's longevity record. But these pioneering studies indicated that krill, in fact, make quite good experimental subjects—if you treat them right.

The experiments on krill growth and shrinkage in the laboratory highlighted the difficulties of determining how old any individual krill is. Length, the standard measure of age, was of little value because of their ability to shrink, but it had the virtue of being a simple measure. Sophisticated analysis software can be used to tease year classes out from the length measurements of a krill population, but uncertainty remains. These findings thirty years ago spawned research programs that are continuing to this day, using a bizarre array of techniques in attempts to improve on our traditional measures of age. None of the new methods for measuring age has yet been universally adopted, and most of them are cumbersome and slow going. Thus we are left with length as our prime measure of age structure, but we now know to interpret our results with a considerable degree of caution.

Growing krill from eggs in the laboratory became possible once we learned how to persuade krill to reproduce in captivity. Since they were first discovered it was obvious that adult krill were either male or female. The big males are quite distinctive (to the trained eye of the krill biologist) and are long, streamlined, powerful swimmers. They produce sperm in packets (called spermatophores), which they secrete through special glands between their rearmost feeding legs. They also have elaborate handlike structures on their forward two swimming legs which they presumably use to transfer the sperm packets to the females. I say presumably because, although the mating sequence has now been filmed, the process of transfer has never been seen in sufficient detail to determine what is being done by which organ. As the females mature, they too become more distinctive. Their body swells as their internal ovaries develop until they become decidedly curvaceous. Females have a gland on their underside onto which males deposit their packets of sperms, and, in females that are about to spawn, there are often multiple spermatophores dangling beneath them like a bunch of bananas—evidence of considerable promiscuity in the krill population. The eggs are fertilized as they are spawned.

Studies from early in the twentieth century by the first known female krill biologist, Helen Bargmann, showed that spawning female krill eject their fertilized eggs into the open ocean. The eggs are heavier than water and begin to sink, beginning a month-long journey that sees them leaving the surface as one-celled embryos and returning as multicelled larvae. As they sink they start the laborious process of dividing and developing. After a week, at a depth of two kilometers (1.24 miles), they hatch—the tiny larvae breaking out of the shell membrane—and then start to swim upward. These larvae are just over a millimeter across, and they have two pairs of spiny limbs, which they use to begin the long swim to the surface. At hatching they have no mouth—they must make their long migration to the surface, fighting gravity all the way, fueled only by the reserves laid down in the egg by their mother. It takes around three weeks of tortuous swimming to reach the sunlit layer of the ocean. On the upward journey, the larvae molt through a series of shape changes, finally developing a rather useful mouth, which allows them to begin feeding when they reach the upper layer of the ocean where algae bloom and food is (hopefully) abundant. By this point they have run out of the fuel bequeathed to them by their mother so they need to begin feeding immediately. Unlike the adults, if there is little food where they emerge they quickly starve to death. Because the spawning season is during high summer (December–February), these larvae reach the surface when it is getting toward autumn and the Antarctic growing season is coming to a rapid and frigid end. They now have an adult-like tail and are bigger—around three millimeters in length—but they still have no fuel reserves, so how do they find enough food to see themselves through winter?

As we have seen in chapter two, most of the krill habitat is covered by pack ice in winter. Finding out what happens in the ocean during the long Antarctic winter was almost impossible until the 1970s, when research icebreakers first began to penetrate the pack ice in this dark and dangerous season. We normally think of krill as free-swimming creatures who spend most of their lives in open water, away from the surface and far from the ocean bottom. Observers on the first winter scientific voyages reported that the ships would turn over fractured floes of ice, which would reveal wriggling krill on their underside, as well as the intense ice discoloration caused by the ice algal community. Krill and their larvae, it turns out, were feeding on these pastures at a time of year when there were few other sources of food in the water. The vulnerable larvae disappeared under the winter ice with virtually empty fuel tanks, and emerged the following spring as developed juveniles.

Before a krill larva develops into an adult it grows through a bewildering variety of different stages—up to nine in number—changing shape with each transitional molt. It is not until they become juveniles that they are quite obviously krill, and by this time they are quite large—around two centimeters in length—and can swim independently. With all these different larval stages, all dependent on similar food resources, it would make sense if there were some segregation of the young from the adults, and the eggs and larvae from all the larger stages. Adult krill are voracious feeders and occur in vast swarms that can clear the water of most particulate material. This is no place for a krill egg or larva to linger. It appears that the large prespawning females migrate to deeper water to lay their eggs. This has two advantages. First, the sinking eggs will hatch before they reach the ocean floor where there is a host of animals waiting to devour anything that sinks out of the sunlit layer. Second, by laying their eggs in offshore waters away from the bulk of the adult population, they keep them clear of the most likely cause of their early demise—their fellow krill. This life history has evolved so that krill can exploit all the conditions in their Antarctic habitat and can cope with the variation that occurs around the continent as well as between years and seasons. Other species of krill in the Antarctic and worldwide have quite different life histories. Some species lay eggs that float, and others brood their eggs until they hatch, so the evolution of a sinking egg—and the spawning behavior that accompanies it—is an adaptation that has allowed Antarctic krill to become so phenomenally successful. What we know about the adult life history is more complicated still.

Krill quite obviously have a complex life history that involves vertical and horizontal migrations, feeding in the upper ocean for some of the year and on the ice algae at other times, while utilizing the drift of the currents and ice to keep them in productive waters. From early studies onward it was observed that adult krill were more abundant near the surface during the night, and they spent their days in darker waters at depth, reaching the surface at sunset and remaining there until dawn. This phenomenon is known as diurnal vertical migration, and it is exhibited by many groups of marine animals. There are numerous advantages to such behavior. By remaining mostly in the dark, krill minimize predation by animals that hunt by sight. By being constantly on the move, krill can also better utilize all the available food resources. The various layers of the upper ocean all move in different directions and at different speeds. If a krill leaves the surface and dives to two hundred meters (219 yards) for twelve hours, then reemerges in the surface layer, it will be in a different patch of water from the one it left the previous morning. This allows a krill swarm to deplete the water of algae and other organisms during the night, descend to depths (where they can also find good sources of food) for the day, then swim back up into a pristine oceanic pasture, which they can exploit during the next night. If all else fails, the seafloor is also a food resource that krill can exploit, and we are beginning to realize that this happens regularly. These vertical excursions expose krill to massive pressure differences. The pressure in the ocean changes rapidly with depth—one atmosphere for every ten meters (eleven yards) of depth. So at two thousand meters (two thousand two hundred yards) the pressure is two hundred times that at the surface. We suspect that such pressure affects the physiological processes within the body of krill, but this is difficult to study in the laboratory, and this means that there is always some doubt over the realism of our findings because the experimental conditions are far from the actual environment.

Krill are also capable of complex horizontal migrations—we know they move offshore to spawn, but recent studies suggest the entire krill population may migrate inshore to overwinter in underwater canyons and deep bays. Although it is difficult to observe whole populations of krill moving en masse across the ocean, there are observations of individual krill swarms actively moving through the water. I have seen such behavior myself, but others have described it far more beautifully than I could:

They were all swimming hard and going round and round, sometimes in a circular course, and sometimes in a "figure 8," but never breaking away from one mass. The cloud would sometimes change shape, elongate in this way, or that (There appeared to be some guiding "principle"—almost as if there was some leader in command of the whole!). (Marr, 1962, 154, quoting descriptions of swarms of krill observed by Alistair Hardy off the jetty at Grytviken, South Georgia, in the 1930s)

Krill are quite clearly capable of complex individual and group behavior, but it was not until krill were kept in aquaria that we started to understand just how complex the social life of krill can be. Anyone who has seen krill alive can describe their active swimming behavior. They dart across the tank, swim both up and down, in figures-of-eight, spiral around purposefully and spend some of the time sinking slowly using their feathery swimming legs as mini-parachutes to slow their descent. Some lucky people have observed krill swimming in schools where they act in perfect harmony with their neighbors, keeping a fixed distance but reacting to the movements of their schoolmates with beautifully executed synchronized maneuvers. Schooling behavior has proved difficult to generate in an aquarium, but it does happen when a suitable density of animals is held in the right light conditions.

In the wild, krill schools move through the water with all the constituent members facing the same direction and swimming actively. When krill are still densely packed but not swimming in the same direction, or when they are forming a confusing melee of individuals, their behavior is described as swarming. I suspect krill alternate between both types of aggregation depending on whether they are actively moving or are more stationary. Certainly, when they are actively feeding, their regimented schooling ranks break up into a swirling mass of animals more intent on gorging themselves than maintaining their social position.

Krill can detect and react to the water currents generated by the rhythmic beating of the swimming legs of their schoolmates, and this is one of the mechanisms they use to stay together. Their big eyes must also play a role. As they spend so much time in the dark, they may also use luminescence to keep in touch with their neighbors. All species of krill have a series of light organs that line the underside of the abdomen and a pair near the genital openings and behind the eyes. One of the most striking features of newly collected krill is the light show they put on. Each animal emits a series of flashes of electric blue light, some of very short duration and others lasting for up to thirty seconds.

This spontaneous bioluminescence becomes less frequent the longer the animals are kept in the laboratory, and after a while a stimulus must be applied to persuade the krill to flash, such as a photographic flash or certain chemicals. In one experiment, a particularly effective chemical that caused continuous luminescence in North Atlantic krill was LSD (this research was carried out in the 1960s). The downward-looking light organs might be used to hide the silhouette that krill must make when seen from below against a well-lit ocean. This behavior would protect them from visual predators lurking below. The position of other photophores suggests that they must have some behavioral function, particularly for reproductive activity, but because observations of krill luminescing in the ocean are so rare, and the behavior disappears in captive krill, we have to resort to speculation.

Some of the best descriptions of swarming and schooling behavior of krill have come from scuba divers who have been fortunate and hardy enough to enter the water to swim with krill. They have described krill schools as curtain-like, or forming huge hemispherical, dome-shaped aggregations. The best depiction of such krill schools I have seen is not in the scientific literature but in the film *Happy Feet 2*, where huge swirling clouds of krill dominate the screen.

Krill in aggregations have been observed to react to predators in unison, forming a halo of empty water around the intruding animal—or diver. If the predator gets too close, then the krill react with rapid backward movements—tail flipping or "lobstering," which can set this behavior off in a wave among their neighbors. Krill in captivity react to perceived threats too. My colleague So Kawaguchi, who developed our krill aquarium into a world-class research facility, has a fine head of dark hair. When he peers into the aquarium holding tank, the krill scatter. This is not because of the cruel and unusual punishment he exerts on the inhabitants, it is because of the contrast between his dark hair and the white environment of the laboratory.

When I loom over the tank, the krill are far less perturbed. I'd like to think that they detect my benevolent feelings, but in truth it is probably because my shock of white hair provides a less alarming contrast.

There has been much discussion in the biological literature about the causes and advantages of aggregating, a behavior that occurs in a vast number of animal groups from whales to mosquitoes. Living in huge, dense swarms gives some protection—safety in numbers. If your swarm is targeted by a predator, then the odds that you will end up as a meal are smaller if you are surrounded by millions of your relatives. On the other hand, if a swarm is discovered by a bulk feeder, such as a baleen whale, then both you and all your neighbors will end up contributing to this year's blubber accumulation. Krill-feeding whales depend on finding large, dense aggregations of krill and could not have evolved without them. A tongue-in-cheek calculation suggested that if krill occurred at average densities, rather than in highly dense patches separated by open water, baleen whales would have to travel at about the speed of sound to catch enough. This has not been observed, so we must assume that whales evolved to exploit aggregated prey and that their feeding behavior and the swarming behavior of krill developed in parallel, with advantages to both groups.

The size of krill aggregations can be staggering. One concentration of krill found off the Antarctic Peninsula covered one hundred square kilometers (thirty-nine square miles) (the size of a small city) and contained two million metric tons of krill. This swarm was targeted by around three hundred humpback whales, which seems, at first glance, like a large number. But, given their maximum feeding rate the whales could only have eaten a minuscule 0.007 percent of the total swarm biomass. Therein lies one of the advantages of aggregating—if predators manage to find your swarm, they cannot possibly wreak havoc on it the way they would with a smaller, bite-sized, aggregation.

We don't know how common these large "super-swarms" are. They are, on purpose, hard to find so they only show up occasionally on our surveys. I have seen massive aggregations only twice in Antarctic waters, and we could only detect them from the records on the echosounders—they were entirely below the surface—but they extended for as much as twenty kilometers in one direction. The surface of the ocean was eerily calm, no feeding frenzy by diving birds, no disturbance caused by foraging seals, and only a few minke whales casually eating their fill at their leisure. Yet these aggregations contained a vast portion of the region's krill—the aggregating behavior seemed to be working. By concentrating so much of the biomass in a small area the krill were making it much harder for the predators to find them.

Living in dense aggregations has disadvantages too. Krill are very oxygen-hungry so large numbers of krill rapidly deplete the water of this vital element. The elaborate aggregation shapes described by divers might ensure that oxygen concentrations never fall too low to affect the constituent krill. Krill also excrete ammonium as a waste product of their ravenous feeding. A build-up of ammonium can render the water toxic. So krill swarms must keep moving to access clean water.

It has been suggested that these polluting effects of swarms effectively limit their dimensions and overall size, but they still manage to reach sizes that few other species can achieve. Feeding aggregations of krill produce a rain of solid waste matter as well, and, believe it or not, scientists, including myself, have spent uncomfortable hours in cold laboratories dropping krill turds and shells into a column of seawater to measure how fast they sink. This is not idle curiosity; sinking solid krill waste is thought to be a major route for transporting carbon out of the surface layer and into the ocean depths. But much of the solid waste may never make it into deep water. I imagine that if you live in a swarm there is a good chance that you may end up eating your neighbor's feces or cast-off shell. Being hungry and omnivorous has its downsides, but there is no advantage to being a picky eater when you live in the ocean's largest aggregations.

The habit of aggregating into swarms or schools appears to be a fundamental, almost defining, behavior of most species of krill throughout the world. Despite this the phenomenon is little studied—other than through attempts to understand the acoustic returns on echosounders. Sure, this behavior is difficult to study, but understanding how, when, and where krill aggregate is essential to determining how the entire Antarctic ecosystem works.

In a strange twist, while krill scientists have largely sidestepped the study of krill behavior, computer programmers have adopted the behavior of krill in swarms as the very model of group behavior. The Krill Herd Algorithm is used to efficiently solve a wide range of optimization problems. It is based on the simulation of the herding behavior of krill individuals. Ironically, scientists studying krill rely on computer models of passive particle flow to describe krill behavior, but computer programmers use detailed behavioral models to describe physical processes.

Considerable strides in understanding the biology of krill have been made over the last forty years, particularly once techniques were developed to keep them alive in aquaria. These studies continue to produce results that send krill scientists back to the drawing board, scratching their heads. But, the development of successful techniques for collecting krill and keeping them alive in laboratories has opened up a host of areas for krill research. Krill have now been reared through seven generations in aquaria—eggs have been grown through all the larval stages, have reached maturity, have mated and spawned, and the offspring have gone through the cycle another six times.

This achievement would have been unthinkable fifty years ago. We can now investigate the long-term effects of environmental changes, such as ocean acidification, over successive generations. We no longer have to travel to Antarctica to study krill; they are available all year round in much more convenient locations. We even know how to manipulate their reproductive cycle so that they will produce eggs when we need them to, not according to a timetable set when they were living in the Antarctic. The whole field of biological rhythms is now open for study; how does their biochemistry change seasonally, what triggers reproductive behavior, when do krill engage in schooling behavior, and when are they more independent?

Our knowledge of krill started with astute observations and beautiful descriptions but soon got sidelined into the dull routine of oceanographic sampling. Our newfound ability to study living krill, both in the laboratory and in the wild, will hopefully drive our studies to new heights of understanding—and a greater aesthetic appreciation. The techniques for keeping krill alive in aquaria make it possible for the public to see living krill and appreciate their captivating behavior.

Hopefully, greater public exposure will lead to the extinction of the phrase uttered by all first-time krill viewers, "I didn't realize they were so big!" By viewing and appreciating krill as complex living organisms we will get closer to understanding the greater complexities of the Southern Ocean ecosystem.

*Chapter Five*

# Antarctic Fast Food

An account of the food chain in Antarctic waters that omitted
any mention of krill would be equivalent to an account of
Hamlet without the Prince of Denmark.
—Louis Joseph Halle, *The Sea and the Ice:*
*A Naturalist in Antarctica*

Antarctic krill has, somewhat understandably, been outcompeted for public attention by the larger and more obvious Southern Ocean species. These animals, often referred to as the charismatic megafauna, are the icons of the region—seals, whales, and penguins and other seabirds. This attention has not always been benign; two centuries of slaughter preceded our current conservation focus in the Antarctic. Meanwhile, krill have remained obscure—despite being the preferred food for these larger (and cuddlier) animals.

There is no doubt about it—penguins are cute, though possibly not as cuddly as is generally imagined. An individual Adélie, waddling across the ice curiously inspecting human visitors, gives rise to all sorts of anthropomorphisms. Such encounters far out into the pack ice can also give a false impression of the personal hygiene of penguins—they appear immaculate in a gleaming white and black livery.

But then they have just had a seawater rinse and an ice scrubbing. Visit penguins at home in their colony and many of these initially favorable impressions disappear in a total sensory overload.

Penguin colonies are unbelievably noisy, with adults and chicks constantly making their presence known in highly discordant tones. The stench of partially digested, slowly rotting seafood is overwhelming, and on close inspection, the colonies resemble the aftermath of a battle rather than a harmonious commune. The rocks are littered with decomposing and dismembered bodies of adults and chicks overlying an older sedimentary layer of the bleached bones of their ancestors. Sinister, predatory skuas swoop from the high ground to pick off unsuspecting chicks or injured adults. And over all exposed surfaces lies a thick, gooey, pink blancmange—a pungent carpeting of krill-rich guano. On the surrounding snow, the nutrients from the colonies stimulate the growth of microscopic algae that stain the once pristine whiteness red, yellow, and green. Adjacent moss banks are stained pink, and the summer melt streams carry the nutrient-rich slurry into the ocean, prompting colorful algal blooms.

Krill from far out at sea are the driving force behind the production of these shore-based ecosystems. Seen from the air, penguin colonies resemble pink oases on the overwhelming brown of the tiny areas of exposed rock around the Antarctic coastline. The color of the colonies changes in response to the availability of krill. When krill are scarce, the penguins bring back more fish, and this stains the rocks white. When krill are abundant, the rocks can be a dark maroon color. These vistas of krill-colored penguin colonies are a visible indication of the importance of krill in the Antarctic food web. But from the krill's point of view, penguins are voracious predators, each consuming up to five kilograms (eleven pounds) a day. There are thirty-four million penguins breeding in or around the Antarctic so their requirement for krill is huge. A century ago, however, they were relatively minor players in the krill-based ecosystem.

To the modern visitor to the Antarctic region it is difficult to comprehend how different the Southern Ocean must have looked a hundred years ago. Accounts of the abundance of whales seem almost mythical. An account from 1892 sets the scene: "Whales' backs and blasts were seen at close intervals from quite near the ship, and from horizon to horizon . . . the sea was swarming with *Euphausia*." Most of these whales would have been the largest animal that has ever lived, the blue whale.

Antarctic blue whales can weigh up to 180 metric tons—over ten times the weight of an elephant—and they get that big by feeding exclusively on krill. Female blue whales reach sexual maturity at about ten years and can potentially give birth to a calf weighing two and a half metric tons every two to three years. The suckling calf can put on weight at the mind-boggling rate of ninety kilograms (two hundred pounds, equivalent to the weight of an adult human) every day. They can live for up to ninety years. This extraordinary rate of production becomes more astonishing when you consider that they are only in the Antarctic feeding grounds for half the year, where they dine on krill at the rate of around six metric tons a day—the equivalent consumption to 1,200 penguins. Once they leave the Southern Ocean for warmer waters in autumn they are thought to starve and live off the vast amounts of blubber they have accumulated during the southern summer. In their heyday, there may have been about two hundred thousand of these giants making the annual pilgrimage to the Southern Ocean to feast on the poor persecuted krill population.

Blue whales are baleen whales. Instead of having teeth they have a series of horny plates that protrude from their upper jaw. These baleen plates were prized for corset making in the nineteenth century. But baleen plates have a more fundamental function than in the world of Victorian lingerie. Baleen whales have an enormous mouth, and the underside of their jaw consists of a ribbed muscular throat that can be extended like a balloon. To feed, the whale opens its massive mouth and engulfs a swarm of krill, then its tongue, the size of an elephant, is used to push the captured eighty metric tons of water out through the series of hairy fringes on the baleen plates, retaining the wriggling krill, which are then swallowed. From the air, feeding blue whales look like lunging tadpoles, their long muscular tail being dwarfed by their hydraulically inflated head. Each mouthful dispatches a metric ton or two of krill.

Blue whales were not alone in their annual krill-fueled journeys.

Fin whales, almost as big and numerous as the blues, and the smaller humpbacks both took their summer vacations in Antarctic waters, where they joined the much smaller minke whales, which may tarry longer in the deep south. In total, whales alone are estimated to have consumed over two hundred million metric tons of krill a year—before they were hunted almost to the point of extinction.

Whaling in the early part of the twentieth century was ruthlessly efficient. Populations of the slower humpback whales were targeted first, but as vessels became faster the populations of fin and blue whales were systematically hunted down and wiped out. This seventy-year orgy of killing occurred on factory ships and in shore-based factories. On the island of South Georgia, the ruins of some of these six factories remain, rusting slowly on the shores of majestic fjords that once ran red with the blood of the giants of the deep.

Walking around these rusting slaughterhouses I was astonished by their sheer size. Huge oil tanks rust alongside the boilers, slipways, and cranes that permitted this industrial-scale slaughter at the edge of the world. Whale catchers list alongside rotting jetties, and discarded harpoons litter the site. The phrase "dark satanic mills" entered my mind as I wandered through these deserted monuments to unchecked greed. These are massive factories, akin to steel mills or car plants, that have been transplanted to this windswept island at the end of the Earth with only one lethal purpose. No one was ever going to remove them—the prey would be destroyed, the profit taken, and the ruins would remain. One million three hundred thousand great whales were killed in the Southern Ocean in less than a hundred years.

The factories spawned their own ecosystems—the carrion feeders that thrived on the gory business of dismembering whale carcasses. In the 1970s concerns were expressed that the cessation of whaling might cause great damage to the populations of petrels that had grown accustomed to the easy pickings of the whaling industry. On land, rats and mice thrived, and imported reindeer took over the mountains. But these changes in the food webs near the whaling stations were minor compared to the wholescale alteration of the marine ecosystem.

The ruins of the Grytviken whaling station on the island of South Georgia. A massive whale jawbone is seen in the foreground. (Photo by author)

It is inconceivable that the removal of the dominant predators in the Southern Ocean ecosystem occurred with no collateral damage. Did the krill population surge in the absence of predation? Did other krill-eating predators thrive on this bounty of food? Or was there a more subtle and unexpected change?

We only have partial answers to these questions. We know that populations of baleen whales (and earlier, fur seals) had crashed because human activities were responsible for their demise, and we have reasonable documentation of their decline in numbers. Records from whaling and sealing expeditions reveal the increasing difficulties that were faced in searching for viable numbers to hunt. Sealing ended when there were almost no fur seals left to catch, commercial whaling stopped when the populations were at between one and ten percent of their preexploitation levels and a moratorium was declared.

We know that both the seals and the whales depended on krill and that they must have consumed almost unimaginable quantities of their preferred food every year. But in the 1970s, ecologists began asking what had happened to the ecosystem when the main predators of krill were removed. The simple answer put forward was that the krill population must have exploded, thus providing a bounty for the remaining predators and later for the embryonic krill fishery. Unfortunately, life (and ecosystems) are not that simple.

The most likely recipients of the "krill surplus," as it became known, were already abundant krill-eating species, such as the seals, penguins, and minke whales. The unfortunate aspect of the krill surplus hypothesis is that we do not have any good data that can be used to test it. We know how many whales were killed and consequently how much krill they would have eaten—a staggering one hundred fifty million metric tons a year. But, our knowledge of changes in the rest of the ecosystem is much shakier.

In the early twentieth century there was no attempt to systematically monitor the Antarctic ecosystem to see if any of the components were changing radically in response to the reduced whale populations and the suggested increased krill abundance. The Discovery Expeditions in the 1920s and '30s focused on krill but provided us with little quantitative data that could be used to assess how abundant they were back then.

There are other historical datasets available for krill, but prior to the 1970s there was no methodical approach to estimating krill abundance, and the data we have has many drawbacks. Although some brave studies have attempted to piece together a picture of historical change from such data there is a great lack of certainty surrounding the methods and the results. We know that the krill population in an area can vary massively from one year to the next so disentangling long-term signals from patchy data is fraught with uncertainties.

Perhaps the best indicator of whether the krill population has changed is to examine the populations of the animals that are dependent on them, which are often easier to count—surely if massive changes have occurred in such a critical ecological link as krill, then these changes would flow through to their predators.

Seals are abundant in the Antarctic region, and two large species are highly dependent on krill. Fur seals have experienced massive human-induced population changes over the last two hundred years. Adults weigh between fifty and two hundred kilograms (110 and 440 pounds), and their cute appearance belies an aggressive nature. Many a seal researcher bears the scars of an encounter with a belligerent fur seal, and even the frisky, inquisitive pups have been known to nip and harry scientists. Fur seals are found mainly on Antarctic and sub-Antarctic islands and the Antarctic Peninsula and were prized for their pelts, which were used to make hats. They were hunted to the brink of extinction in the eighteenth and nineteenth centuries and have taken most of the twentieth century to slowly recover in some of their habitat. There were estimated to be around four million breeding fur seals in 2015, though it is difficult to gauge their numbers very accurately. Fur seals eat krill so their recovery may have been aided by the demise of the great whales, but they are probably not even at their preexploitation level so their recovery is not conclusive proof of an increase in availability of their prey.

Crabeater seals are less well known than fur seals, probably because they live in the far south and have never been the target of commercial exploitation. Crabeaters are reckoned to be the most abundant seal in the world and are large, weighing up to three hundred kilograms (660 pounds). Despite their name, they are specialized krill-eaters that live out on the pack ice year-round. They eat about twenty kilograms (forty-four pounds) of krill a day, swimming in packs in the waters offshore of the sea ice, porpoising through the water and diving for their prey, even down to the ocean floor. When not feeding, they can be found lounging on ice floes.

One of the great pleasures of traveling through the pack ice in an icebreaker is standing at the bow, far away from the hum and throb of the engines, absorbing the experience of the vessel cracking and moving aside great slabs of ice. Startled seals are a familiar sight. Crabeaters seem to be easily ambushed by large, noisy icebreakers and are seen looking up in bewilderment at the last minute when the largest, reddest object they have ever seen disturbs their slumber.

They too soil their pure-white resting places, and their krill-based feces are often mistaken for the bloody evidence of conflict and carnage. Crabeaters' numbers have been difficult to estimate but are thought to have remained stable or may even have increased over the last century. But the small increase suggested is not enough to make a significant dent in the proposed "krill surplus."

Penguin numbers can be more easily monitored because they return to colonies each summer and can be counted. Scientists are now using aerial and satellite imagery to estimate how many colonies and penguins there are along the Antarctic coastline in summer, and we now have far more accurate data than we have ever had. Early information on penguin numbers is scarce, and it was not systematically collected so it is difficult to detect long-term trends. Overall, it appears that penguin numbers have changed, though not in a concerted fashion, and some populations are increasing, whereas others are decreasing. Again, the changes detected are not sufficient to mop up the "krill surplus."

The most contentious potential beneficiary of the "krill surplus" is the minke whale. Like the blues, fins, and humpbacks this is a baleen or filter-feeding whale, but it is much smaller. This small size meant that it was exploited much later, and its numbers were never greatly reduced. Because of its abundance, there is still interest in catching this species commercially, but there is currently a moratorium on commercial whaling. This has led some nations to argue that minke whales have increased massively because of the "krill surplus" and need to be culled to promote the recovery of the larger species of whales.

Counting whales in the open ocean is fraught with difficulty and results in abundance estimates that have large errors; thus it has been very hard to collect data and analyze them to detect any trends in the minke population. Certainly, there is no evidence that their population underwent an explosion when their competitors were removed, but this does not stop whaling under the guise of "scientific research" being conducted to bolster the case for a resumption of full-scale commercial whaling.

There is little convincing evidence that any of the remaining major krill consumers—seals, seabirds, or minke whales—increased dramatically in numbers following the demise of the great whales. Similarly, there is a lack of information on changes in the community of less visible animals—fish, squid, and other invertebrates—in the Southern Ocean. For fifty years, there has been a great deal of speculation over the fate of the "krill surplus," but what if there never was one?

The krill surplus hypothesis rests on the assumption that the major effect of large animals is through their consumption of animals and plants lower down the food chain, but the major effect of the great whales might not have been what went into their voluminous mouths but what came out their other end. To understand how this could possibly be we must delve into the processes that make oceanic ecosystems thrive.

Life in the ocean is very different from life on land. On land, plants are big, and the animals that eat plants can range from the smallest forms of terrestrial life to the largest. In contrast, in the ocean, most of the plants are unicellular, suspended in the upper layer near sunlight, and their grazers are generally at the smaller end of the size spectrum.

The largest animals in the ocean are carnivores, whereas in terrestrial environments the largest animals are herbivores. Krill are about as big as oceanic herbivores get.

The bite-sized krill can be easily seen by animals that hunt visually, and the extensive and dense krill swarms can be exploited by large, bulk-feeding predators, such as whales. So basically, every type of large animal in the Southern Ocean can and does eat krill. On the flip side of the coin, krill are sufficiently large and mobile that they can catch and eat almost anything smaller than themselves, and this includes most of the living (and dead) material in the ocean. This ecological constriction means that the energy from the sun that flows through the Southern Ocean ecosystem up to the larger animals must pass through krill. This is what ecologists, for obvious reasons, call a wasp-waist ecosystem, and this funneling of energy and nutrients through a single species makes krill uniquely important in the Southern Ocean food web.

There are other marine ecosystems where single species play an equally dominant role, but the Southern Ocean is probably the largest expanse ruled over by one species. Krill are often described as being at the "base of the Antarctic food-web," which entirely misrepresents their central ecological role and instead relegates them to the position of phytoplankton—the true base of the ecological pyramid.

The term *phytoplankton* is a catchall for an extensive variety of photosynthetic microorganisms. The diversity found in the phytoplankton is astonishing, with small, motile flagellates, diatoms enclosed in silica shells, cocolithophores housed in spherical containers made of circular plates of calcium carbonate, and primitive cyanobacteria. About all they have in common is that they are unicellular and hence microscopic in size, and they photosynthesize. Because of their diversity, they vary greatly in shape and size, from the smallest at only a few micrometers to chains of diatoms that are so large they can be seen by the naked eye. Despite their small size, they are responsible for nearly all the ocean's productivity and most of the world's oxygen production.

The microbial ocean world does not stop at the primary producers. There is a whole universe of unicellular organisms, bacteria and viruses, that prey on phytoplankton or attach themselves to dead and decaying material, reducing organic matter back into the chemicals from which it was formed. What this means is that most of the ocean's life is invisible to us, but not to the animals that form the next link in the food chain.

Krill are by no means the only herbivore in the Southern Ocean, but they are probably the best studied. The frantic lifestyle of krill means that they need a lot of fuel, and for most of the year they must obtain this from the suspended matter that surrounds them in the ocean. The front five legs of a krill have a whole series of fine hairs down their entire length. When the legs are extended, the hairs connect between them and form what is known as the feeding basket. The gaps between the hairs are truly tiny (~ 0.005 millimeter [~ 0.0002 inch]), and the insides of the legs are also lined with fine-meshed spines. The entire apparatus is designed to expand, capture a volume of water with its constituent food particles, then compress, expelling the water and keeping the food, which is then moved toward the mouth. This method of feeding is called filter feeding and is remarkably similar to the method used by the major predators of krill—the baleen whales.

Once the particles are brought to the mouth they are crushed by the horny mandibles and then ingested to the stomach, which is well muscled and has another series of filters and spines. The stomach's function is much like that of a bird's gizzard—to further physically break down the ingested food before it is passed on to the larger digestive gland, where it is digested. This barrel-shaped organ, clearly visible in living krill, is the home of powerful enzymes that rapidly break down the food. Krill have a suite of enzymes that is unique in its ability to break down protein rapidly, and as we shall see in chapter six, this has both positives and negatives for the krill fishing industry.

The food of krill has traditionally been thought of as mainly diatoms. These one-celled plants live inside a shell made of glass-like silica, and the tough mandibles of krill are adapted to break open these cells to enable access the goodies inside. Recent research has shown that, although krill rely on diatoms when they are abundant, they will also eat virtually anything else in the water. This is not surprising because when you live with several trillion members of your own family you cannot afford to be picky about what you eat.

But plants are only really productive in summer, and krill have to survive through the long, dark winter when algae are scarce in the ocean. We know that krill can shrink if food is scarce, and this allows them to conserve energy at the same time as fueling their bodily needs from their own internal reserves. But is food really scarce? Tantalizing glimpses of the winter environment have come from the few winter voyages into the krill habitat, and these suggest that food, in certain areas, may not be in short supply. There are the pastures of ice algae, though this is thought to be more important for young krill, and there is evidence that krill become less selective in their feeding habits during winter, eating more animals and detritus than in summer.

Krill may eat up to twenty percent of their body weight a day. This represents a massive amount of food that krill need to cram in just to keep themselves going, growing, and reproducing. They probably spend a large part of each day feeding—when there is food around. Krill have this great need for energy because they must keep swimming all the time. With all this food going in there must be a similarly large amount of waste matter being produced—and there is. Feeding krill are often trailed by long strings of fecal matter that eventually break off and sink.

A number of dedicated scientists have gone to some considerable length to collect these krill turds and measure their rate of sinking. The motivation behind this seemingly perverse activity is because the waste material of krill, and other swimming animals, contains a range of chemicals that the animals do not need. These chemicals include carbon-based products, and the process of krill feeding takes some of the carbon that has been taken up by growing plants that are floating in the surface waters and transforms it into rapidly sinking waste material. In deep water the organic material decays, and the chemicals are remineralized back into the water by microorganisms, such as bacteria. This is one of the mechanisms that rapidly transports carbon out of surface waters to the depths of the ocean and to the seafloor.

Historically, the primary oceanographic role that animals were seen to fulfill consisted of eating algae, thereby consuming the carbon and other elements the algae had taken up through photosynthesis, and then converting the algae and those elements into fast-sinking fecal material and other particulate waste materials. This process of transferring organic matter and nutrients swiftly into deep water works for all levels in the food web. It also applies to the cast shells of pelagic animals like krill, and to the enormous bodies of dead whales.

It is becoming apparent, however, that in addition to direct consumption and transfer, there are other, more subtle, ways of affecting the cycle of carbon and other elements in the marine system. Animals have specific needs for individual elements and can accumulate these minerals (either from their food or directly from the seawater) in ways that are species-specific. If there are enough of those animals, and they have a need for a particular element, then they have the potential to affect the way in which that element is cycled through the ocean system. Krill have a peculiar need for fluoride, a chemical that is very scarce in seawater. They incorporate it into their shells, and this has caused headaches for the krill fishing industry, whose products contain too much fluoride to be used to feed humans or domestic land animals. Krill lose much of their fluoride every time they molt, and they then rapidly take up fluoride from the seawater to replace the lost ion. The regular molting of krill must be one of the major determinants in the distribution of this element in the ocean.

Another vital element in the ocean is iron. Iron is a trace element that is essential for phytoplankton production and growth. In large parts of the world's oceans, particularly the deep open ocean areas, there is not enough iron in the water so the growth of phytoplankton is limited. There can be so little iron dissolved in seawater that it is difficult to quantify. With the advent of sensitive new instruments, and clean collection and analysis techniques, the measurement of iron in seawater has become routinely possible. Experiments have shown that adding soluble iron to the surface waters in these iron-poor areas results in an almost instantaneous bloom of phytoplankton.

The Southern Ocean is one of the vast oceanic areas where phytoplankton growth is limited because of the lack of dissolved iron in the surface waters. Iron is hugely abundant in the Earth's crust and enters the ocean through wind-blown dust, from rivers or from the seafloor, and through upwelling of deeper water that has become enriched in nutrients through the breakdown of organic material sinking out of the surface layer. Antarctica lacks rivers and is mostly ice covered, so it is a poor source of windborne dust, and the continental landmass contributes little iron to the Southern Ocean. Iron levels in the surface layer of the Southern Ocean are so low that any mechanism that maintains this element in the surface layer and allows it to be recycled would help to maintain productivity. This may be where whales and krill come in.

In the early 2000s, a colleague of mine controversially suggested that whales could fertilize the surface layer of the Southern Ocean by recycling the iron found in their prey. The basic concept is that large adult whales, when on their feeding grounds in the Southern Ocean, are consuming large quantities of iron-rich krill, but because they are producing blubber (which contains little iron) rather than muscle (which is high in iron) they have little need for the iron in their diet. Whales, like most mammals, have no mechanism for getting rid of excess iron from their bodies once they have absorbed it so they only take in the iron that they need—any more would make them ill. In humans, having too much iron in the body causes the disease hemochromatosis. Consequently, when adult whales are feeding, the iron in their food passes straight through the digestive tract and exits the body in a plume of fecal material, which may act like liquid manure. Thus was born the "whale poo hypothesis," and it was greeted with intense skepticism by the scientific community, myself included.

Despite my initial doubts, a team of us decided to examine the fundamentals of the whale poo hypothesis by measuring the iron content of krill and of whale feces. The obvious first question that is always asked is, how do you collect whale poo? The answer is, with a very long-handled net! Baleen whales always defecate at the surface, and their waste products come out as an oily, red, evil-smelling slick. These slicks can be mistaken from a distance for surface swarms of krill, but on closer inspection their unmistakable stench gives them away. Characteristically, they are a mixture of thick liquid and lumpy solids, which makes them difficult to sample, but which probably results in them being rapidly dispersed, thus making the nutrients rapidly available to the rest of the ecosystem.

At my institute in Australia, we had been persuading researchers and whale-watching operators to collect samples of whale feces whenever they came across defecating whales, so we had amassed the world's largest collection of whale poo, which we stored in a convenient freezer. This proctological exercise was originally devised to examine the feces using modern genetic techniques to determine what the whales had been eating. This additional information was a bonus to our studies—we could use these samples to examine how much iron they contained, and we could also tell what the whales had been eating. We could also obtain samples of the species the whales had eaten and determine their predigested iron content. Our first discovery was that concentrations of iron in whale feces were at least 10 million times the background level in seawater and, not surprisingly, the flesh of whales and the krill they had been feeding on also contain similarly high concentrations of iron.

Some simple sums drove us to conclude that around twenty-four percent of the iron in the top two hundred meters (219 yards) of the water was tied up in krill bodies. Because krill are strong swimmers, they can keep this significant quantity of iron suspended in the upper layer of the ocean, but it is in their bodies—not available to the phytoplankton. Because krill are long-lived, they can keep all that iron in the surface layer throughout the year and from year to year, with the population acting as a buoyant reservoir of iron. This is where whales enter the story. By eating krill and rejecting most of their ingested iron, whales convert the reservoir of iron present in the bodies of krill into liquid manure, which ought to stimulate phytoplankton growth. Experiments continue to see whether whale effluent has a fertilizing influence, but we feel that our first attempts to refute the hypothesis had failed. Whale poo might well be ecologically important, and this importance would have been greater in times past.

Unfortunately, there are relatively few baleen whales in the Southern Ocean now, but this fertilizing effect of krill and whales would have been far greater when whales were abundant. Great whales were estimated to consume 150 million metric tons of krill a year before industrial whaling, so they would have been converting most of the iron present in the krill population into iron-rich feces readily available for phytoplankton growth. If whales and krill do (or did) play a major role in recycling iron in the upper ocean, then this explains how an ecosystem with more animals in it can also be a more productive ecosystem. It also suggests a solution to the krill surplus hypothesis. When humans removed the fertilizing influence of whales, the Southern Ocean ecosystem lost the ability to be so productive. Fewer whales meant fewer phytoplankton and krill, and so less food for other krill-consuming animals. Thus my initial skepticism has been overturned, and there is now a vibrant community of researchers looking at how various species of animals positively affect the productivity of the ocean.

It seems unlikely that the role of animals as buoyant nutrient reservoirs is restricted to the Southern Ocean, or solely to the iron cycle. In the Southern Ocean, air-breathing vertebrates have a major ecological role, and this is probably because this region lacks a prolonged history of systematic exploitation—most of the overexploitation is relatively recent. Examination of historical records and descriptions leads to the conclusion that all the world's oceans used to sustain large numbers of seals, whales, and seabirds, as well as vast populations of large fish and reptile species. The Southern Ocean was the last region to be exploited by hunters and fishermen, and the ecosystem they saw there, rich in large animals, probably resembled northern oceans a thousand years ago. The ecosystems of the Pacific, Atlantic, and Indian Oceans are probably only a shadow of their former selves and used to teem with large animals—before exploitation changed them forever because the loss of the large animals led to a decline in ecosystem productivity.

There are other ways in which animals can influence the productivity of the oceans, particularly by encouraging the mixing of nutrients from the deep into the surface waters and by bringing up nutrient-rich water and sediment from the seafloor. Some calculations have suggested that the daily vertical movements of animals may be one of the main forces mixing the oceans and beneficially redistributing nutrients equivalent in magnitude to the effect of winds and tides.

What's more, these calculations are based on estimates of the current density and abundance of swimming animals. Because animals, particularly larger animals, were far more abundant in the past, the effect back then would almost certainly have been much greater.

These effects of animals on their environment have been termed ecosystem engineering. Research proceeds apace to examine the combined effects of animal movements and nutrient storage and recycling. It is already apparent that the simplistic arguments suggesting that you need to cull large predators to increase fish (and krill) yields are not credible, and large animals, including krill, play critical roles in structuring ecosystems. In the Southern Ocean, we are in the middle of an experiment. The great whales are recovering, some species at an astonishing rate, and we are conducting long-term ecological research in several regions. The recovery of the great whales will be a very strong ecological signal that will allow us to examine our concepts of ocean ecology. Will more whales result in more krill? I hope so.

Although scientists now acknowledge that there are many distinct food webs in Antarctic waters, after a century of study we still accept that Antarctic krill dominates the ecosystem that feeds the large vertebrates of the region. Coinciding with the demise of many of the large krill-feeders, and spurred on by talk of a "krill surplus," a new and, up until now insatiable, predator has arrived in Antarctic waters—humans.

# Chapter Six

# Eating Krill

I believe, then, that the cod fishery, the herring fishery, the
pilchard fishery, the mackerel fishery, and probably all the
great sea fisheries, are inexhaustible.
—T. H. Huxley, Inaugural Address to the 1983
Fisheries Exhibition in London

I am sitting on the bridge of the *Antarctic Sea*, which lies marooned
in the dry dock in Montevideo undergoing corrective surgery
to turn it into the world's largest and most sophisticated krill
fishing vessel. The deck is a jumble of cables, hoses, scaffolding, and
discarded slabs of metal. In the vast hold, giant chunks of machinery
lie haphazardly in the cavernous space that one day will be occupied
by metric tons of processed krill. In the distant forecastle, a maze
of stainless steel pipes, conduits, and conveyors are being assembled
into the processing plant. This floating factory will produce what the
Norwegian fishing company Aker Biomarine hopes will be a range
of products that will justify the huge investment in their three-vessel
fleet. Even without the chaos of the refit, the ship is not a pretty
sight. The captain jokes that the *Antarctic Sea* had been voted one of

the world's ugliest ships several years running, and it is easy to see why. It is long, looks ungainly, and is quite unlike any fishing vessel I have ever seen. It is missing the usual trawl deck, nets, and winches associated with conventional fishing practices. The *Antarctic Sea* used to be a freighter. It was converted for catching and processing in the early 2000s for a Norwegian fishing company that subsequently went out of business—a not unfamiliar fate for many krill fishing operations. Aker Biomarine bought the 133-meter (145-yard) vessel and commenced a costly refit in Uruguay before the vessel was re-born as the 9,500-metric ton *Antarctic Sea*. The vessel had originally been named *Thorsovdi*, Norwegian for the "Hammer of Thor," not a particularly sensitive name for a company whose stated vision is "to create a healthier lifestyle through responsible and sustainable decisions." But what might justify such massive investment? and why would anyone want to go fishing for krill in the first place?

Although the Antarctic region is often referred to as a pristine ecosystem, we know that this is far from the case. The first expeditions to the Southern Ocean were commercially motivated, and these pioneers were looking for seals, whales, or other species that they might be able to harvest. When the populations of seals and then whales were exhausted, extractive industries looked elsewhere. Although there are early records of fish being caught, it quickly became apparent that the Antarctic region lacks the vast populations of the north, such as cod, herring, and anchovies. Fish have been caught commercially in the Southern Ocean since the 1970s, and small fisheries continue today. There were, however, krill, and thoughts soon turned toward using krill as a resource. There had been earlier attempts to make use of this obvious resource; a German expedition in the early 1900s reported dryly that they tried eating krill:

> We caught the creatures [krill] in such quantities that we were able to eat them too; they tasted quite good, but were rather small and tiresome to peel. (Erich Dagobert Von Drygalski, *The Southern Ice-Continent: Reports of the German South Polar Expedition 1901–1903*, reprinted by Bluntisham Books [Cambridgeshire, UK: Bluntisham, 1989], 132)

It took another seventy years before the world's fishing industry took any real notice of the vast stocks of krill in the Southern Ocean, but the qualities of abundance, size, and peeling difficulty have remained with the krill fishing industry up to the present time.

Commercial harvesting of Antarctic krill took a while to become a reality. When it became apparent that whaling in the Southern Ocean had come to a bloody conclusion because most of the populations of whales had been driven to the point of extinction, fishing nations began looking for other Antarctic species to exploit. Fur and elephant seals had been exterminated in most of their colonies in the 1800s, and in 1972 a treaty was signed to regulate any future harvesting of seals—there has been little serious attempt to revive this industry in the life of this treaty. Penguin harvesting for oil had been forcibly ended as a result of one of the first wildlife conservation campaigns in the early 1900s. That left fish, squid, and of course, krill. Fish were caught in quite large numbers in the late 1970s, but then it appeared that these were slow-growing species that did not replenish themselves quickly so the catches declined rapidly and subsequent landings have remained low at less than twenty thousand metric tons a year for much of the last forty years. In comparison, northern fisheries that provide for the table (e.g., cod, haddock, pollock, and herring) around the world often catch hundreds of thousands to millions of metric tons annually. For many reasons, Antarctic waters have never been rich in fish, and after the early exploitation efforts they became less so. Squid are thought to be abundant in the Southern Ocean because so many animals, such as elephant seals, albatrosses, and sperm whales, obviously eat a lot of them. Peculiarly, neither scientific research voyages nor fishing expeditions have been able to catch significant quantities of squid, which has led to the conclusion that there may be a lot of squid in the Southern Ocean but they are rather shy and are fast enough to avoid most nets. So that left krill.

Following discussions of the "krill surplus," which was theoretically the legacy of the removal of the great whales, in the 1970s Antarctic krill became lauded as the next great marine resource that was destined to provide protein for the world's hungry. It didn't quite turn out that way.

The first ships to investigate the krill resource were those from the Soviet Union and Eastern Bloc countries—significantly, these were enterprises that were not required to operate on a conventional economic model. The massive costs of operating a fishery in Antarctic waters were not borne by the companies exploiting the resource. Japanese vessels were also early entrants in the krill fishery, and, despite the dubious economic returns available, they remained fishing for krill until 2011, when Japanese companies decided that it was cheaper to buy krill that other people had caught. This was an era when fishing fleets were expanding and were exploring even more distant waters in search of new stocks. Populations of conventional, and more coastal, species were being depleted rapidly, and fishing companies sought out new, and more distant, fishing grounds and species. Penetrating the Southern Ocean required large vessels and usually fleets of ships because of the difficult conditions and extreme distances from ports.

Earlier chapters have emphasized the large size of krill, but it is time to admit that, in terms of commercially harvested species, krill are rather on the small side. There are other species of small shrimps that are fished and consumed in other parts of the world—usually by artisanal-scale coastal operations. There are also inshore fisheries for species of krill in the North Pacific around Japan and in British Columbia. But the Antarctic krill fishery was the first attempt to develop a distant water fishery on a species of this nature. Operating in the Antarctic region was difficult enough, but developing the fishing technology to deal with bulk catches of krill was a major challenge. The first decade of the fishery saw vessels investigating catching strategies and companies feverishly developing catching and processing technologies and exploring the products that could be produced from krill.

Early on it became apparent that the curious biology of krill would begin to intrude into human attempts to make anything of value from their vast population. A curiosity that had been noted by early biologists was that krill began to turn black rapidly after death. Scientists solve that problem by pickling krill in formalin, an option not available to the fishing industry, which needed an edible product. Krill, as we have seen, need to eat ravenously during spring and summer, and they need the digestive system to deal with a massive input of protein that has to be converted into swimming energy, body mass, and reproductive output. It turns out that a suite of murderously effective digestive enzymes has evolved inside krill. These enzymes can reduce krill food to soluble protein in short order, but they can also turn on a krill's own body once it dies. Fishermen noted that krill begin to turn black after only an hour on deck, and within three hours they are so autodigested that they were useless for producing any desirable item. This rapid degradation occurs at a near freezing point, which illustrates the power of these proteolytic enzymes. The rapid spoiling of the krill catch was frustrating for fishermen. Because krill swarms are so dense and so large it is possible to catch large quantities in a very short period of time. A conventional trawler can land ten metric tons an hour and can process ten catches a day. The rapid spoilage of the catch meant that fishing captains had to exercise extreme restraint and could only catch the tonnage of krill that they could process in a couple of hours. Slow processing meant that fast profits were unlikely.

No sooner had the fishing industry become used to catching krill in a fairly relaxed fashion, than krill threw another wrench in the processing works. In 1979 a couple of Norwegian researchers decided to examine the fluoride content of krill. They were surprised—the fluoride content of whole krill was very high indeed—about the same concentration as is found in fluoridated toothpaste. Brushing our teeth twice a day with a high-fluoride concoction might be good for our teeth, but it is quite another matter to ingest large quantities of fluoride in our food. Fluoride is an essential element and helps vertebrates, like us, to develop hard bones and teeth, but too much results in discoloration and destruction of our much-needed hard parts. Unfortunately, the amount of fluoride in krill put them well into the danger zone for most terrestrial vertebrates. A major rethink was required in terms of how krill were processed and the types of products that the industry was developing. Papers produced in the late 1970s by scientists working in the Soviet Union indicate some of the problems. Much of the Soviet catch was being used to feed domestic animals, such as pigs and chickens. The papers pointed out several structural and developmental problems suffered by domestic animals raised on krill meal. The discovery of the high fluoride levels in krill provided an explanation for these problems, and the industry had to change what it did with its captured krill. A large drop in the catches in the mid-1980s has been partly attributed to the industry regrouping to change its product mix.

Pigs, cattle, and chickens rarely eat krill in the wild so they lack the enzymatic pathways to deal with ingesting more fluoride than their body requires. Their bones and teeth tend to accumulate it to the point where it becomes toxic. This is not the case for many marine vertebrates, some of which consume large quantities of krill as part of their natural diet. Whales, seals, and many species of fish can eat large quantities of krill (of all species), accumulating fluoride in their bones without harm, while their flesh remains essentially fluoride-free. Thus developing products using krill for fish food was an obvious way forward for the fishery.

There has been a concerted effort to produce low-fluoride products so that the market for the krill catch could be wider than just as aquaculture feed. The initial results of fluoride analysis of krill were derived from homogenized whole krill. The result was an average high fluoride level, but it disguised the fact that the shell was fluoride-rich, whereas the rest of the animal was fluoride-poor. Get rid of the shells and the resultant muscle is quite acceptable for consumption by people or domestic animals.

Producing a low-fluoride krill product requires either removing the shell or removing the fluoride from the shell. We don't yet know the chemical form that fluoride takes in the shell of krill, but we suspect that it plays a role in the hardening of the shell after molting. A newly molted krill rapidly takes up fluoride from the seawater and stores it in its shell. Unfortunately, the fluoride in the shell appears to be very mobile and highly water soluble, and once the krill dies this troublesome ion begins to migrate rapidly into the muscle. For the fishery, this means that peeling must commence rapidly on capture, and only freshly caught krill can be used to produce peeled krill products.

But how do you peel krill? There are literally dozens of patented methods for peeling krill (involving a frightening array of rotating drums, whirring blades, and jets of air or water), but such complex processes merely increase costs and generate waste. Because the fluoride is water soluble there ought to be ways of washing it out of the catch. There are several suggested methodologies for doing this—but again all this extra processing increases costs and slows down production. Nonetheless, the fishing industry has persisted, and krill have been continuously harvested since the 1970s, though the products and the major players have changed with time.

Over the years, as the krill fishery developed its catching technology, annual catches of krill declined from the high points of the early 1980s, with regular landings of two to three hundred thousand metric tons per year currently being reported. There has been limited concern that the overall level of catches is unsustainable since the estimated biomass of the krill population at the start of the fishery was around half a billion metric tons. Besides, there has been a comprehensive treaty in place to regulate the fishery since 1982 (The Commission for the Conservation of Antarctic Marine Living Resources, or CCAMLR as it is known, which we will explore in the next chapter).

Up until 1993, most of the krill was being caught by large fleets of supertrawlers from the Soviet Union. These fleets comprised up to twenty-five fishing boats, tankers, scouting ships, research vessels, and

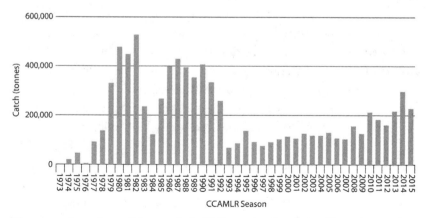

The annual catches of Antarctic krill from the Southern Ocean since the fishery commenced in the early 1970s.

refrigerated transport ships. They fished right around the Antarctic continent and operated without any real scrutiny. The catches were annually reported to the Food and Agriculture Organization of the United Nations and, since the early 1980s, to the international management body (CCAMLR), and some additional scientific information was also reported, but by and large, the fishery was shrouded in secrecy.

In the early 1990s I, alongside my colleagues working in CCAMLR, attended two krill fishery–related meetings in the Soviet Union. This was a rare opportunity to find out more about what went on in the Soviet krill fishery and to meet with scientists and technicians who worked with krill. In the margins of these meetings our hosts made much of their development of high-quality products for human consumption. They fed us banquets featuring peeled krill tails, showed us factories producing krill paste, and gave us souvenir cans of peeled krill to take home—I still have some of mine, but it is probably way past its use by date. We were given a cookery book that we could use when we opened our cans of krill, which included tempting recipes, such as krill meat in Dutch sauce, croquettes of krill meat, and cucumbers stuffed with krill meat. The krill products they fed us were gastronomically unremarkable. It turns out that the population of the Soviet Union felt rather the same way, and, despite food shortages, it proved difficult to move krill products off supermarket shelves.

But this was only part of the story. It appears that, despite the propaganda, the vast bulk of krill caught by the Soviet Union was not destined for human consumption. I have already mentioned the use of krill for domestic animal use, but krill were also used as food for carnivorous animals, such as mink. There are even disturbing reports that tens of thousands of metric tons of krill were plowed into fields as fertilizer—a practice that indicated the lack of economic accountability in the Soviet system. Unsurprisingly, as the Soviet Union began to disintegrate, and as fishing companies began to have to operate on a basis of providing an economic return for their huge costs, the Soviet fishery declined. Russia and Ukraine continued to dabble in the krill fishery (and still do), but other nations took over the reins of krill fishery development.

Production of krill paste in the Yalta fish factory in the old Soviet Union. (a) Krill concentrate was mixed with butter and then (b) extruded into containers. The resultant pink product was not particularly attractive, had an unremarkable taste, and smelled decidedly marine. (Photos by author)

Since the Second World War, Japan has been a major fishing nation, and its fleets have spread to all the oceans of the world, including the Southern Ocean surrounding Antarctica. The Japanese krill fishing fleet was less organized than that of the Soviet Union, and several different companies sent individual vessels or small fleets of two or three vessels to Antarctica to develop the krill resource. This practice continued from the late 1970s until 2011 and Japanese vessels continued to catch between fifty and one hundred thousand metric tons of krill a year, although, as their experience grew, the number of vessels declined. The Japanese industry and government were not as secretive as the Soviet Union had been, but it was still difficult to obtain reliable information on the operation of their krill fishery and on the products emerging from it. Part of this difficulty was because of the supposed commercial nature of such information, but a large part was also that the world's fishing industry at the time had a culture of revealing little unless forced to do so.

Fishing for krill was not a new phenomenon for the Japanese. A fishery for North Pacific krill, known as *isada*, began in the 1940s, using bow-mounted nets targeting annual swarms of krill on the sea surface in the Sanriku area. This fishery persists until today, and it is an important inshore fishery. Two other species of krill are similarly harvested in the Japanese coastal zone, and the total catch of all species around Japan has occasionally exceeded one hundred thousand metric tons a year. Most of this is used to feed farmed fish, and it is a prized source of aquaculture nutrition because krill contain red pigments that naturally color the flesh of species such as sea bream and salmon. So the Japanese fishing industry had developed some techniques for dealing with krill and had some ready markets for Antarctic krill.

In the 1990s and early 2000s the krill fishery in the Antarctic settled down and consistently caught just over one hundred thousand metric tons a year, most of it being landed by Japanese vessels. The economics of the fishery remained shaky, thus preventing significant new investment, but other nations also fished for small quantities of krill; some, such as the South Koreans, consistently, others, including the United States, Chile, and Poland, more sporadically. Many nations continued research and development on krill with an eye toward producing something that might make the huge cost of krill fishing profitable.

The developmental phases of the krill fishery can be tracked by looking at the range of patents that companies have taken out over time for processes and products. Companies only lodge patents for a product or process if they think they can make money out of it and if they think they have an idea worth protecting. A staggering 812 krill-related patents were lodged between 1976 and 2009. In the early phase of the krill fishery many patents were lodged for methods of catching krill effectively. This was followed by a phase when the industry concentrated on novel ways of producing food items and animal feed. Then came aquaculture, and finally there is a recent emphasis on pharmaceuticals and other biochemical products.

The recent move from food and feed products to pharmaceuticals and neutraceuticals (biochemical health foods) is an interesting trend because it signifies a shift from bulk products to the production of smaller quantities of high-value products, such as krill oil.

There is no doubt that Japanese efforts in the 1990s and early 2000s were concentrating on producing edible and profitable products. I own several lavishly produced Japanese brochures from the 1990s with titles (in English) proclaiming "Krill—Healthy Cooking," "Krill Scene," or "The Rich Harvest from the Antarctic Ocean to Your Table." These brochures provide recipes, alongside photographs of models in swimsuits, for such tempting dishes as krill cannelloni, krill egg rolls, fried krill in chili sauce, stew with krill meatballs. These recipes used peeled krill in either tinned or frozen form. This glossy literature was obviously being used to educate the Japanese public (and because some of the brochures were in English, other consumers too) about the nature of krill, known as *okiami* in Japanese, and its potential culinary uses. Japanese companies also produced krill sausages and fermented sauces, and considerable resources were devoted to inventing new and acceptable food items from krill. Despite the fluoride issue, there was also a market for frozen or dried whole krill for human consumption.

As a human, I have conducted several experiments on consuming krill. Krill, not surprisingly, tastes distinctly shrimplike, though the females that are ready to spawn have a rich crabby flavor. These flavors are slightly offset by the texture when one is eating whole krill. There is a crunchy, fibrous texture that will be familiar to anyone who has eaten small shrimps that are served unpeeled in many parts of the world. The high-fiber element can also wreak havoc with the digestive system if too many krill are consumed—not recommended because of the fluoride issue anyway. Canned peeled krill tails have a slightly sulfurous odor but otherwise taste much like minced prawn or shrimp meat. The peeled krill tails are small when cooked so they resemble pinkish grains of puffed rice—fine as ingredients but not of much use as a stand-alone food item. Krill have been used as an additive to provide a shrimplike flavor to a range of processed foods.

My poor children got used to me bringing home krill-containing food items that I had discovered in my travels or at the supermarket, and they became my expert tasting panel. They were reluctant to eat the smelly krill directly from the tin, but they gave a thumbs-up to products that ranged from Japanese krill sausages to Australian processed fish sticks, which revealed their Antarctic contents in the fine print on the side of the packet. The krilly ingredient of these products became apparent when the occasional black eyeball was encountered staring out from the breaded interior. The difficulty of removing the eyeballs was, in fact, a major impediment to the acceptance of krill as a food item. A colleague of mine spent months on a Japanese krill trawler bent over the conveyor belt with a pair of tweezers at the end of the peeling line meticulously picking eyeballs out of the constant flow of krill tails. (This is not the way to produce a cost-effective end-product!) The massive increase in the production of farmed shrimps and prawns means that they are now far cheaper and more available than they were thirty years ago, which means that krill cannot compete with them on price or attractiveness to the consumer.

Even though there was a public emphasis on the use of krill to feed people it is likely that most of the Japanese catch was used as aquaculture feed, aquarium food, or, surprisingly, sports fishing bait. Many species of oceanic fish feed naturally on krill of various species. It is this diet that makes salmon and sea-run trout pink, and because it is a dominant food source of wild fish it is also close to being nutritionally perfect. There has been considerable research into producing aquaculture feed based on or at least including krill, particularly for vulnerable life-history stages of species such as salmon, when providing the right nutrition is critical. Krill also have the real advantage of being naturally pigmented with astaxanthin—the substance that turns the flesh of salmon pink or red. Most farmed salmon use synthetic astaxanthin, so farmed fish that obtain their coloration from eating krill can be marketed for a premium price because they are "naturally colored." Catching Antarctic krill and processing them to be fed to farmed fish is expensive, and frozen whole krill are bulky, so most of the krill catch these days is reduced to meal. This dried meal has most of the same properties as whole krill, but it can be compacted considerably. It is, however, costly to produce, and, despite its undoubtedly beneficial properties, it is far too expensive to compete with fish meal from more conventional species. Consequently, krill meal is used sparingly and is marketed as a specialized product that attracts a premium price.

Other approaches to producing aquaculture feeds from krill have involved enzymatic digestion of the krill under controlled conditions, often using the krill's own powerful digestive enzymes. These processes produce either a liquid or a dried powder, and low-fluoride feeds can also be produced this way. Small quantities of these hydrolysates have been used as feeding stimulants in the diets of farmed fish, and the addition of even small amounts (two percent) can make previously unpalatable food sources, such as soya meal, attractive.

The Japanese fishing industry expended huge amounts of effort and money on sending ships to the Antarctic to catch krill. A surprising amount (approximately thirty-four percent) of this krill was brought back to Japan and then thrown straight back into the water. Krill are a prized bait for sports fishing and are also used as chum to attract fish for anglers to catch. There were dispensing machines in Japanese coastal towns that produced packages of krill for this use, and considerable effort went into marketing different types of products for a variety of other uses. There was even one type specially developed for women, with a built-in scoop for the krill, because it was thought that Japanese women were unwilling to use their hands to remove the krill from the plastic package. The krill were chemically treated to ensure they remained intact when used as bait, and this also made them resistant to decay. Apparently, there are large parts of Tokyo harbor that are biologically dead, covered in a layer of slowly rotting krill thrown there by anglers eager to attract fish. Ironically, krill from pristine Antarctic waters end up contributing to the fouling of one of the world's most polluted waterways.

Biochemists also became interested in the powerful suite of enzymes found in krill that had so frustrated commercial fishermen. Proteolytic enzymes have a range of uses in medicine and industry, and the obvious power of the krill enzymes made them a target for research. Initial research focused on their ability to break down flesh—medically useful when removing decaying tissue from recovering burns and wounds. Researchers conducting trials found krill enzymes to be highly successful at cleaning up recovering lesions, but there was an interesting side effect—the wounds rarely became infected. It appears that krill enzymes also have an antibiotic property. This finding spawned a whole new field of research, and patents were subsequently lodged claiming the amazing medical efficacy of krill enzyme preparations. One such patent claims that the product can cure a bizarre range of conditions and diseases, including athlete's foot, foreskin infections, eye infections, gum infections, common colds, influenza, bronchitis, gastric ulcers, herpesvirus, acne, eczema, dental plaque, cancer, HIV, hemorrhoids, thrombosis, cataracts, glaucoma, wrinkles, warts, diarrhea, baldness, and leprosy. Other patents have been more modest in their claims but have suggested that krill enzymes can be used to clean dead and living material, such as textiles, hair, furs, skins, plastics, leather, nails, ears, mirrors, glass, porcelain, denture prostheses, metals, stones, teeth, facades, works of art (such as paintings), and even contact lenses. Interestingly, the enzymes found in Antarctic krill are unlike those found in other species and are difficult, if not impossible, to clone artificially. This means that the only known source of these enzymes is from Antarctica. Extracting commercial quantities of these enzymes from krill is tricky, and, because they are likely to be used in small quantities, the tonnage of krill necessary is correspondingly small. Krill enzymes might one day be an interesting and lucrative by-product of the krill fishery, but alone they will never justify the expense of establishing a krill fishing venture.

There are other products and chemicals that can be extracted from krill; some of these are very valuable, but none of them has yet made its way out of the laboratory or off the patent lawyer's desk and into the marketplace. An obvious chemical that can be extracted from krill is chitin—the material from which the shell is made and a waste product of peeling krill. This valuable chemical (and its derivative chitosan) has a wide range of uses, including water purification, manufacturing of biodegradable plastics, use as artificial skin, and as a propriety fat-binding dietary aid. Take several chitosan tablets before a meal, so the marketing literature suggests, and you can gorge on fatty foods, which will exit the body bound to the chitosan, and you will remain slim. Unfortunately, chitosan also binds to several essential vitamins and minerals so there may be undesirable side effects.

If producing krill meat by peeling krill becomes a commercial reality, then large quantities of chitin will be produced as a waste product. The question then becomes whether it is cost-effective, and safe, to process this material on the ship and transport it to factories on shore where it can be used. Chitin is roughly 2.5 percent of the dry weight of krill so a two-hundred-thousand-metric ton-per-year fishery produces only about one thousand metric tons of chitin. Other sources of chitin include prawns, shrimps, crabs, and even fungi so supplies are potentially huge. It seems unlikely that anyone would go fishing for krill just to extract chitin, but it might become a valuable by-product.

The economics of the fishery have been shrouded in mystery right from the time the first fishing vessels set out to catch krill. There are very few records of what was produced in those early years, and even fewer indications are available for what those catches were worth. Either the value of the fishery was viewed as a state secret or it was claimed that such data were "commercial in confidence." Recently, with the advent of a more open and transparent approach to the fishery, more data are available on the wholesale value of krill products. The estimated average value of all products from the krill fishery is approximately US$69.5 million per year over the last five years. Krill meal is available for approximately US$70 per metric ton, which makes it far more expensive than other fish meal produced from anchovies or sardines. Krill oil, however, is a very valuable product, selling wholesale for about US$21,000 per metric ton. Only a small fraction (approximately ten percent) of the krill caught are processed for oil, and if the demand for this product increases, then it is entirely possible to produce far more krill oil without catching more krill. Flooding the market with krill oil would drive down the price and possibly make fishing for krill an entirely uneconomical activity. Both the meal and the oil will have to be niche-marketed as premium products if the fishery is to make a profit—this suggests that the fishery cannot grow too large, or too fast.

Perhaps the most significant development in the krill fishery to occur since the demise of the Soviet Union was the entry of Norwegian companies into the krill fishery in 2009. Norwegian involvement brought with it not just new technology and products but a new attitude. Throughout the history of the fishery there had been an antagonistic relationship between the fishing nations and the management body established to regulate the krill fishery. This meant that information on how the fishery operated, the economics of the operations, or the types of products was only reluctantly and incompletely provided. The biggest fishery in Antarctic waters was shrouded in mystery despite being subject to considerable scrutiny. Within two years of entering the fishery, Norway became the largest krill fishing nation in Antarctic waters and the dominant force in the industry. The attitude of both the Norwegian government and the Norwegian companies was a breath of fresh air to the krill fishery. It almost seemed that greater transparency and a willingness to cooperate in the management process were calculated trade-offs for sustained long-term krill catches.

The Norwegian companies have concentrated on two products: krill meal and krill oil. Krill meal was destined primarily for aquaculture, and Norway, as a leader in marine aquaculture, had a natural market. Krill oil, on the other hand, was a developing market that was riding on the back of a massive body of research that has demonstrated the value of omega-3 oils for a whole range of disorders, from high blood pressure to attention deficit/hyperactivity disorder. Krill oil contains a mixture of different types of fats or oils, including omega-3s and phospholipids, and thus can deliver many of the proven benefits of consuming marine oil supplements. But like krill meal, krill oil is more expensive to produce than most other sources of omega-3 oils, largely because of the difficulties of fishing in Antarctic waters. The way around this is to conduct research to determine whether krill oil is equal to or better than other marine oils. This research has, to date, produced equivocal results, but there are other differences between krill oil and its competitors. Most notable among these differences is that krill oil is less likely to produce reflux—studies have shown that consumers are prepared to pay more for oil capsules that do not make them burp. This one factor, more than any perceived medical or health benefits, is the key marketing advantage that krill oil has. There is, however, considerable research into the medical effectiveness of krill oil and into the production of drugs based on krill oils, so there will be a gradual move out of the health food arena and into mainstream medicine. The market for omega-3 oils is dynamic, and new players are constantly emerging; for example, calamari oil is now available in addition to more conventional fish oils—so maintaining a toehold on supermarket shelves requires constant vigilance and an emphasis on research and development. Ensuring that the costs of producing krill oil are as low as possible has also required the development of new technology.

This brings us back to the unconventional Norwegian krill-fishing vessel, the *Antarctic Sea*. The ship is a converted freighter, not a trawler like most of the vessels that fish for krill, but with the recently developed technology there is no need to trawl. The *Antarctic Sea* and its sister ship the *Saga Sea* use a sophisticated pumping technology to transport the unwitting krill from their swarm seamlessly into the processing factory. The patented system (Eco-harvesting Technology®) utilizes a small net, which is carefully positioned into a krill swarm that has been detected using sophisticated sonars. Air is then injected into the end of the net (the cod end) and the air and a parcel of water containing krill is lifted up a hose and onto the ship. This is referred to as continuous fishing.

Despite the name, the Eco-harvesting fishing technique has been greeted with horror by the conservation movement with dramatic headlines such as "Hoovering up the base of the food chain" being used to decry this practice. Although it sounds ecosystem-threatening, the pumping technology does have several advantages over conventional trawl. Krill fishing has traditionally used huge pelagic trawls, capable of catching a minimum of ten metric tons per hour. Often, because of the density of krill swarms, this catch level is achieved within minutes. These nets are towed through the water and may be damaging significant amounts of marine life that encounter the wings of the net. One of the problems with any conventional trawl is that the net's contents are a mystery until the catch is landed on deck. If there is a significant amount of bycatch, or if what is landed is not the expected species, or if the catch is more than the factory can process, then it is too late, and the unwanted catch is discarded.

Conventional trawling can be an incredibly wasteful process. With the pumping technology, the flow of the catch can be regulated to provide just the amount required for processing. The incoming flow can be monitored continuously, and if a bycatch or significant amounts of undesired species are detected, the operation can be halted. Potentially, there is less waste and much less collateral damage from fishing. The catch rates can also be much higher—the pumping ships can land about eight hundred metric tons a day compared to about five hundred metric tons a day for conventional trawlers. Given the regulatory limitations on the fishery, however, this means fewer ships fishing rather than more krill being caught. Other companies are currently investigating the use of similar pumping technologies, including some of the Chinese fishing enterprises.

Future Chinese involvement in the krill fishery is currently the great unknown. Chinese vessels have been fishing for krill since 2010, and catches by Chinese vessels have, at times, climbed to fifty thousand metric tons per year, making them a major player in the fishery. Mixed messages are coming out of China with news of potentially massive investment in krill fishing tempered by reports that the fishing companies have been told that they must make a genuine profit—a difficult task when it comes to krill fishing. So far, the Chinese fishing fleet has been composed of older vessels diverted from other fisheries, but there is talk of a new state-of-the-art vessel that will be able to compete technologically with the Norwegians. Even with this vessel, there is no guarantee of profitability, as the demise of several Norwegian ventures indicates, and the future will lie in producing saleable goods at a reasonable price. Considerable research is being expended in China as part of a five-year development program to investigate the production of commercially viable products from the krill fishery. Papers are emerging on new techniques to remove fluorine and on novel uses for krill, but this is all on the science side of the equation. Time will tell whether the Chinese krill industry can develop the technology to produce commercially viable products for their huge internal markets—or whether they may expand their operations irrespective of the profit motive. Hopefully the active participation of China in the krill management system will ensure that any expansion is within legal limits and is sustainable.

There have been other major changes in the operation of the krill fishery that have not been driven by economic or technological factors. Back in the 1980s it was clear that the pattern of operation of the krill fishing fleets was largely driven by the physical challenges of working in the Southern Ocean. The krill fishery used to operate right around the continent, but in the mid-1990s they abandoned the fishing grounds off East Antarctica and concentrated on the South Atlantic. This area had always been a preferred area to fish because it is the only place in the Southern Ocean where there are significant krill stocks in an ice-free area—the island of South Georgia. The South Atlantic also has a shorter period of the year when the sea is covered in ice compared to parts of East Antarctica, where the fishing season can be as short as four months. These factors meant that the fleets could remain in the South Atlantic year-round, fishing in the south in summer then migrating northward as the ice advanced in autumn.

In the last twenty years, because of the changing climate, the pattern of sea ice advance and retreat has changed. Large areas off the Antarctic Peninsula are now ice-free or have only minimal sea ice in winter, and other areas have a much longer ice-free period. This has meant that krill fishing vessels can operate in this region even in deepest winter. The covering of the krill habitat by sea ice in winter used to give the poor, harassed krill population a much-needed respite from the attentions of predators—including the fishery. Nowadays, predation is a year-round feature, though no studies have yet shown that this has adversely affected the krill population.

This ability to fish in winter has changed the temporal pattern of fishing—the peak of fishing activity used to be in early summer (December–January), but now it has shifted to early winter (May–June). The reasons for this shift are not solely due to changes in the physical environment. Krill are actively feeding on algae in early summer and are often heavily pigmented by deep green chlorophyll. This is problematic for the fishery; products that feature actively feeding krill can be tinged a nauseous shade of green and apparently taste "grassy."The enzyme levels in krill are also highest in early summer so their quality declines more rapidly once they are caught. In autumn, krill contain fewer green pigments and are also richer in the types of oils that are commercially valuable. The movement toward catching higher-quality krill and oil extraction has meant that the peak fishing period now falls later in the year, enabled by greater access due to the reduction in sea ice. A winter fishery also reduces competition between the industry and krill predators, which have a greater requirement for krill when they are rearing their young in summer.

Climate change has favored the krill fishery—and may assist with its management. This extended fishing season offers opportunities for using the fishing fleet to provide the year-round scientific samples that have been difficult to obtain from sporadic scientific surveys. Long-standing problems that have plagued our understanding of krill biology may be accessible using fishing vessels as research platforms. Already we are able to detect annual cycles of growth, biochemistry, and distribution from data collected by the fishing fleet. In the future, most of the information needed to manage the krill fishery may be sourced from the fishery itself.

Understanding where fishing vessels go to make their catches is more in the realm of anthropology than in that of marine biology. As the krill fishery has been operating for forty-odd years there is some historical information that can assist a skipper who is looking for harvestable quantities of krill. But the environment has changed quite dramatically during this period, and there is some evidence that the behavior of krill may also have changed, so maps of fishing locations from the 1980s may no longer be of any use.

In the Antarctic, all vessels come from the north, and they all have a great distance to cover, independent of where they are coming from. South America is the closest continent to regions where krill are abundant, and there are several deep-water fishing ports, such as Montevideo and Punta Arenas, within striking distance of the Antarctic Peninsula. This makes the South Atlantic the logical location for most krill fishing activity, but this is still a huge area with considerable variation, so where does a prospective fishing vessel start?

Scientists and fishermen know that krill populations tend to favor waters just offshore of the continental shelf, but in the Antarctic Peninsula area they are also to be found well up onto the shelf—even to the coastline. Concentrations of krill tend to lurk in undersea canyons, in bays, and in the vicinity of islands. There may be some krill out in the open ocean, but finding it is a challenge, and fishing vessels tend to use the physical cues of bathymetry and shoreline to begin their search.

There are other important factors that come into play in this remote part of the world. Island groups can provide shelter in stormy weather, which is not uncommon in the Southern Ocean at any time of year. Icebergs are an ever-present hazard, and they are more common in some areas than others. As the large glaciers and ice shelves begin to disintegrate more rapidly, icebergs will become more of a factor for the fishing fleet to cope with.

Generally, for a fishing vessel, the goal is not just finding krill, it is more about finding a *quantity* of krill that is worth stopping for and fishing on for a few days. Of course, one of the best clues for locating fishable quantities of krill is the presence of other krill fishing vessels. It seems that, particularly for inexperienced fishing operators, finding out where the rest of the fleet is and staying close to them is a useful strategy. There are also considerable advantages from the point of view of safety in these dangerous waters. Fishing in new and uncharted waters may yield some access to new stocks, but it means that the exploratory fishing vessel is off the radar—literally. There have been cases where Antarctic fishing vessels have run into difficulties, and have even sunk, so being relatively close to other activities is only sensible.

Other activities occurring in the remote South Atlantic include the research and resupply of Antarctic stations, and most of Antarctica's research activities are clustered around the Antarctic Peninsula, for largely the same reasons that the fishing fleets concentrate there— ease of reliable access. The other major activity in the region is Antarctic tourism, a fast-growing and mainly ship-based activity. Every year, thirty thousand tourists pay to visit the Antarctic on cruise ships, and these too favor the Antarctic Peninsula area due to the relatively short (three to four days) sea journey from ports in South America. The area is also largely ice-free in the summer, and there is an abundance of spectacular scenery and wildlife. The presence of so many seals, penguins, and whales results from the same factor that draws the fishing fleet to the region—an abundance of krill.

This leads to a potential conflict. The tourism industry markets the Antarctic as one of the world's last wilderness regions, so the last thing that tourists want to see is huge factory trawlers scooping up large quantities of the food that the wildlife depends on. The tourist operators are aware that their customers prefer to see an untouched environment and go to great lengths to ensure that ship visits are coordinated so that tourist vessels do not encounter each other either at sea or when they make landings at significant sites.

In a similar way, there has been correspondence between the tourist and the fishing industries so that their vessels can try to avoid working in similar areas. From the fishing industry's perspective, this makes sense. Most of the Antarctic tourists are relatively wealthy, educated, and supportive of conservation efforts. This is an influential group, and it also happens to be the demographic that krill oil is marketed toward. Making sure that the operations of the krill fishing fleet are not viewed as intrusive or damaging helps to ensure that Antarctic tourists still buy krill oil capsules when they return to their home countries.

Back issues of fishing industry trade magazines are littered with headlines announcing one company or another is heading south in search of the red gold that is krill. Companies from India, Taiwan, the United States, the United Kingdom, and Australia have all had a good look at the fishery—then their initial enthusiasm has faded away in the face of the harsh economic realities. It is probably fair to say that, in forty years of fishing, no one has yet made a consistent profit from krill fishing. Even the Norwegian krill fishing industry, which has made many long-term and strategic decisions about its krill fishing operations, sports more bankrupt enterprises than successful ones.

Companies continue to fish for krill for a variety of reasons—some political, some because there is a belief that krill will become more valuable in the future, and some just because they want to go krill fishing, and hang the expense. This makes it rather difficult to predict the future of the krill fishery. It seems to me that the krill fishery is highly unlikely to expand to the multimillion metric ton levels predicted in the 1960s and '70s. The environment, both physical and political, has changed dramatically since the fishery first started, so the huge catches predicted earlier seem highly unrealistic now. Even if the economics improve enough to where krill fishing becomes highly profitable, the catch would be limited by the international agreement that manages the fishery in the waters around Antarctica—the Convention on the Conservation of Antarctic Marine Living Resources—known as CCAMLR or "the krill convention." In the next chapter I will explain how an international institution approaches the task of managing the conservation of the ecosystems at the bottom of the world.

# Chapter Seven

# Conventional Approaches

No data, no fish.
—John Heap, UK Commissioner to CCAMLR, 1989

On October 10, 2010, at midnight, the Antarctic Peninsula management zone was closed to krill fishing for the rest of the season. That closure was the first active step to manage the krill fishery. This was the first time in the twenty-nine years of the management regime that there had been a need to take formal action to manage the krill fishery. The fishery was duly closed, the fishing fleets moved on, and the management system for the fishery had passed its first real test.

This is not to say that nothing had been done to manage the krill fishery over the three previous decades; in fact, a raft of measures had been developed that were designed to safeguard both the krill stock and the populations of animals that depended on krill. Most of the management rules are designed to be long term and will only come into effect once (or if) the fishery begins to expand to much higher levels of catch. This approach to management is very different from the way in which other fisheries are managed throughout the world. This novel approach was adopted because of to two factors:

the disastrous history of exploitation in the Antarctic region, and the central role of krill in the ecosystem.

The features that make krill such an obvious target for a fishery also make it fiendishly difficult to manage; krill are hugely abundant and are distributed over a massive area. Critical aspects of their life history are difficult to determine, and their icy, stormy habitat is not conducive to conducting research. Thus developing a system to manage the krill fishery had its difficulties, but these were not entirely biological problems. Back in the late 1970s, when the krill convention was being negotiated, Ronald Reagan was president of the United States, Margaret Thatcher was prime minister of the United Kingdom, and Michael Gorbachev was president of the Soviet Union—none of whom could have been described as an ardent conservationist. Nonetheless, under their respective watches a wide-ranging environmental treaty that covered the waters around Antarctica was rapidly negotiated, and it was focused on the growing krill fishery. How did this happen and why did it happen so quickly?

First, let me issue a warning. Discussing fisheries management can be tedious; it involves the minutiae of biological measurements, bureaucratic processes, and management systems. Further, delving into the intricacies of Antarctic conservation means negotiating a minefield of convoluted acronyms. I could sum up this chapter by writing "The krill fishery is managed by CCAMLR, an ATS-related body that uses the GYM to establish PCLs on the krill fishery, which has been viewed as a major advance on the use of MSY"—but I won't. With the aid of a small glossary, and some devious use of the English language, I hope to forgo the alphabet soup and instead shed some light on the management of the krill fishery.

I have now sat through thirty annual meetings of the Commission for the Conservation of Antarctic Marine Living Resources (CCAMLR, pronounced "camel-are" and hereafter referred to as the Commission) that oversees the conservation and management of the Southern Ocean. I am not alone; there is a cohort of my colleagues who have surpassed the thirty-year mark. These meetings could not be described as exciting, but rather than receiving a well-deserved medal for surviving 420 days of tedium, such veterans are presented with a statuette depicting, not a krill, or even a penguin, but a wombat. To understand the significance of this tangential tradition we must delve into history.

Interest in fishing for krill began to grow in the 1960s and '70s, spurred on by reports of the massive size of the krill population. This was an era when global fishing fleets were expanding, and new fishing grounds were being explored in every ocean. The first forays into the Southern Ocean in the 1960s worried the scientific community. Scientists, more than anyone else, had been active in the Antarctic region for much of the twentieth century, and they had been witness to the ecological damage wrought by waves of sealers and whalers. The prospect of a new era of marine exploitation focused on a vital ecological link spurred the research community into action. Back then, very few scientists working in the Antarctic were researching krill, but they did understand that extractions of krill could greatly affect the animals they really cared about—mostly the seals and penguins. An additional factor ensured that scientists led the way in negotiating a convention that had as its heart the protection of krill—they were the only people who had any idea about what krill were. Ultimately, the lawyers and diplomats negotiated the Convention on the Conservation of Antarctic Marine Living Resources (which, like the Commission, is also known as CCAMLR and is hereafter referred to as the Convention), but its approach and scope came from the scientific community. This made CCAMLR very different from many other international conventions.

In 1976, when it looked like fishing for krill was about to get serious, Antarctic scientists organized a series of meetings under the auspices of groups such as the Food and Agriculture Organization of the United Nations (FAO) and the Scientific Committee on Antarctic Research (SCAR). At these meetings, an international biological research program—Biological Investigations on Marine Antarctic Systems and Stocks (BIOMASS)—was developed to focus attention on the Southern Ocean and the role of krill. BIOMASS involved thirteen countries using thirty-four different vessels in three expeditions. This was the largest biological research program ever mounted, but it was not the first large-scale Antarctic scientific program; the

International Polar Year (IPY) in 1957–58 was a massive scientific undertaking, as part of the International Geophysical Year (IGY). Are you beginning see what I meant about acronyms?

The defining feature of the IPY was scientific cooperation between all nations operating in Antarctica. There were scientific exchanges between nations, and researchers from nations not normally active in Antarctic science became involved. Almost all scientific research in Antarctica is government sponsored, so this huge effort inevitably involved considerable government focus and engagement. A significant outcome of the IPY was the establishment of the Scientific Committee on Antarctic Research (SCAR) in 1958, with a mission to facilitate and coordinate scientific research at a level that was beyond the capabilities of individual nations. In parallel with the formation of SCAR came discussions on the development of a treaty to oversee and regulate all activities in the Antarctic region—the Antarctic Treaty.

Antarctica, the last continent to be explored, has no permanent residents and no agreed-upon territorial boundaries. Seven nations lay claim to parts of the Antarctic. Australia has the largest claim at forty-two percent, and smaller territories are claimed by France, New Zealand, Great Britain, Argentina, Chile, and Norway. These claims are based on many historical and geographic factors, and several actually overlap. In the 1950s there was considerable potential for conflict between countries that claimed similar portions of the Antarctic, and between nations, such as the United States and the Soviet Union, that were involved in geopolitical struggles in other parts of the world.

Significantly, neither the United States nor the Soviet Union had made any territorial claims or recognized any of the claims of other nations to territories in the Antarctic, but they made a point of establishing research stations at strategic positions around the Antarctic. The United States placed a station at the South Pole, in theory occupying territory in virtually every other nation's claim because the claims are pie-shaped wedges that converge at the bottom of the world. The Soviet Union placed several stations in territory claimed by Australia and Norway. There was potential for these polar chess games to get out of hand. There was also concern that the arms race that was proceeding apace in the Northern Hemisphere might spread to the Antarctic continent. Against this backdrop, and given the co-operation of the IPY, the Antarctic Treaty was speedily negotiated and entered into force in 1961.

The really clever aspect of the Antarctic Treaty is that it sidestepped the whole issue of who owned which bit of the frozen continent. The Treaty did not eliminate territorial claims, it suspended them without recognizing any of them. No new claims can be made during the life of the Treaty, and existing claims cannot be amended. Importantly, no activities can affect the status of the pre-Treaty claims. The Treaty applies to all land and ice shelves south of longitude 60° south, and its expressed aim was to set aside the continent for peace and science. It became, essentially, the first arms control treaty of the cold war era when it banned military activity (but not military support) on the continent.

There were several other astute elements to the Treaty. There was to be a system of international inspections to ensure that nations involved in Antarctic research were not cheating. Consultative mechanisms were established, where decision making was by consensus, and there was a review period of thirty years when the Treaty could be open for renegotiation if members agreed.

The first conservation initiative under the Antarctic Treaty System (ATS) dealt with the conservation of plants and animals that lived on the Antarctic continent and islands south of longitude 60° south. This agreement left out those Antarctic animals that lived in the open ocean or in areas north of longitude 60° south. Some redress to this situation occurred in 1978 when a convention to regulate the hunting of Antarctic seals came into force. This convention (the Convention on the Conservation of Antarctic Seals [CCAS]; hereafter the Seals Convention) was developed by the Antarctic Treaty Parties and was made simple by the fact that there had been few serious attempts to hunt seals in the Antarctic region in the twentieth century. This convention was probably most important because it provided the Antarctic Treaty Parties with the experience necessary to negotiate a far more difficult piece of legislation—a convention to regulate fishing in the Antarctic region.

By the time the Seals Convention was signed, fishing in the Southern Ocean was already under way. One and a half million metric tons of fish were caught in the eight years prior to 1978, and krill catches were already topping one hundred thousand metric tons a year. This meant that the Antarctic Treaty Parties, for the first time, had to negotiate a labyrinth of both commercial and political investments if they were to negotiate an agreement.

The science was a bit more straightforward. The British Discovery Investigations in the 1930s had established a good knowledge base on the distribution, abundance, and basic biology of krill. The emerging BIOMASS program was designed to build on this knowledge with the specific aim of determining the size of the krill stock so that the sustainable fishing level could be determined. Research by scientists in the 1970s confirmed the vital importance of krill in the Southern Ocean ecosystem and revealed a few of the biological quirks that make krill such a challenging species to manage (see chapter three). So, when negotiations surrounding the regulation of fishing in the waters around Antarctica in the ATS began, scientists were actively involved in the process.

Whereas the Antarctic Treaty applies only to the area south of longitude 60° south, the new treaty would have to apply to areas much farther north to include the populations of krill and fish living there. The treaty had to apply to international waters as well as to the economic zones of member countries. The waters around the Kerguelen Islands and Heard Island were recognized (by most nations) as being the national territory of France and Australia, respectively. The new treaty would also have to deal with areas whose ownership was disputed, including the waters around South Georgia and the South Orkney Islands, which were claimed by both Argentina and the United Kingdom. This added considerably to the difficulty of the negotiations. The final zone that the treaty would govern was a massive thirty-two million square kilometers (12,355,000 square miles)—over four times the land area of the United States.

There was little consolidated knowledge about the species that would be managed under the new treaty. Seals were already covered by the Seals Convention and whales were being dealt with, somewhat

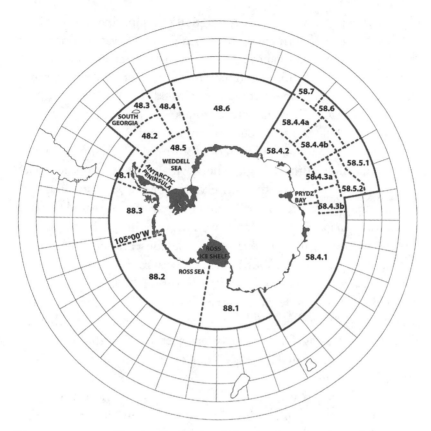

The area managed by the Commission for the Conservation of Antarctic Marine Living Resources, showing the various subdivisions for which krill catch limits have been set. (Drawing by Marcia Rackstraw)

ineffectively, by the International Whaling Convention, but this still left a range of little-understood fish species, many elusive forms of squid and crabs, and poor misunderstood krill. Despite the lack of hard information about krill it was not difficult to convince those tasked with negotiating the new treaty that krill are the foodstuff of all the animals that they, and the general public, were really concerned about. To conserve seals, whales, and penguins required conservation of krill.

Early in the negotiations it became apparent that there would be the potential for conflict between those whose main interest in the

Antarctic ecosystem was the exploitation of its marine resources and those who envisaged the new treaty being a conservation document. Because the discussions about a convention centered around the burgeoning krill fishery this would have to be central to the negotiations, so a pure conservation agreement was ruled out. At the same time the ecosystem concerns that arose from krill fishing, and the parlous state of the whale and seal stocks, meant that any agreement had to be much wider than a standard fisheries convention. Reaching an agreement that allowed for a large fishery with unknown consequences, as well as providing for the conservation of the krill stock and all the animals dependent on krill—and the recovery of the whales and seals, was a major act of diplomacy. In the end, the Convention cunningly referred to its objective being "the conservation of Antarctic marine living resources," but it made a bow to the fishing interests by redefining what was meant by conservation: "For the purposes of this Convention, the term 'conservation' includes rational use." This was a lovely little piece of legal footwork which ensured that the Convention was agreed to swiftly, but even in 2017, thirty-six years later, there is still heated discussion in Commission meetings on what constitutes "rational use."

The Convention was quite specific about its conservation goals, and it laid them out in a series of twenty-three articles and in an introductory preamble. The preamble to the Convention is a delicious entrée to the world of diplomatic nicety. It is a string of paragraphs that sets the scene for the underlying need for a convention. Each of the eleven paragraphs begins with a diplomatic code word: *Recognizing*, *Noting*, *Conscious*, *Considering*, *Believing*, *Recognizing* (again), *Recalling*, *Bearing in mind*, *Believing*, *Recognizing* (one last time) before getting to the ultimate point: *Have agreed as follows*. The Convention, in its twenty-four pages, specifies its scope, ranging from conservation objectives (complicated and innovative), through the official languages (English, French, Spanish, and Russian) to rules of procedure (arcane). I would not recommend reading the text for the Convention (available from https://www.ccamlr.org/https://www.ccamlr.org/en/document/publications/convention-conservation-antarctic-marine-living-resources) unless you are a legal scholar or are suffering from insomnia, but there are several aspects of this Convention that make it unique and are relevant to the krill story, so I will summarize these as briefly as I can.

The Convention established an international commission (also confusingly known as CCAMLR) that meets annually to give effect to the objectives and principles set out in Article II of the Convention. Article II is the meat of the document and still sounds groundbreaking thirty-six years after it was agreed to.

While agreeing that conservation includes rational use, Article II sets out three basic principles. The first principle addresses the requirements for sustainable harvesting of an individual species—the level of population decrease resulting from fishing that is deemed acceptable. This paragraph deals with harvested species in isolation, and is little different from most fisheries agreements of the time, although it is rather more conservative in its aims. But what comes next are the two paragraphs that set the Convention apart in 1982, and which still seem novel now. One paragraph deals with the maintenance of ecological relationships and the restoration of depleted populations. The first point was deliberately aimed at the krill fishery, because of the dominant ecological role of krill in the ecosystem, and the latter point was to ensure that fishing did not affect the recovery of the great whales and the fur seals, both of which depended on krill. The third paragraph of Article II specifies the maximum level of ecosystem disturbance that is acceptable. The aim of this whole series of principles was to be "the sustained conservation of Antarctic marine living resources." This type of conservation agreement ("the ecosystem approach to management") has since been adopted by many other fisheries and conservation bodies throughout the world. The bottom line is that, when any species is harvested, the ecosystem's resilience to the level of harvesting must be taken account. This was a seismic shift in resource conservation.

Paradoxically, while it was being negotiated, and shortly thereafter, CCAMLR was known colloquially as the krill convention, but despite this focus, in its nearly six thousand complicated words, the Convention does not once mention the word *krill*. In a phrase that must have horrified the biologists but which probably satisfied the lawyers, the Convention vaguely stated that it applied to "fin fish, mollusks, crustaceans and all other species of living organisms, including birds." The area of interest was to include all the waters south of the Antarctic convergence, the oceanic boundary that defines the northern limit of the Southern Ocean (see chapter two). This northern boundary was chosen for two main reasons: to include most of the sub-Antarctic islands where fishing had already taken place, and to include all the known krill habitat. So krill was implicitly the focus of the Convention, but it lost its special status when the negotiations moved on from the concerns of scientists (who understood what krill were) to become the playground of lawyers (who really had little knowledge of krill).

Significantly, the Convention also stipulated that the Commission should formulate, adopt, and revise conservation measures based on the "best scientific evidence available." To obtain this information the Convention established a Scientific Committee, a consultative body to the Commission. This reliance on available scientific information has been key to the success of many initiatives within CCAMLR because information on many species is severely limited. Consequently, the advice that the Scientific Committee can offer may be equally limited—but it may still be the best scientific evidence available. It was no longer possible to hide behind the defense that because of a lack of scientific certainty there is no requirement to act. This was especially important in the early years of CCAMLR, when data on all harvested species were lacking, which led to the catch-cry of "no data, no fish," meaning that in the absence of information fishing should be limited. This erring on the side of caution became enshrined in many later environmental treaties as the "precautionary principle." Essentially it is now up to a fishing nation to explain why it is safe to harvest at a given level rather than asking conservation-minded nations to explain why fishing ought to be curtailed at a certain level. This has been termed the reversal of the burden of proof.

The Commission meets annually and discusses issues relevant to the Convention. Because Australia hosted the meeting wherein the negotiations that resulted in the CCAMLR Convention were finalized, it was agreed that the secretariat for the Commission should be

located in Australia. Hosting a secretariat for an international commission is quite a big deal in diplomatic terms, and this was the first
time that such a commission had been assigned to Australia. At the
time of this agreement, most of Australia's marine and Antarctic interests were being relocated to Hobart, the capital of the island state
Tasmania, so the decision was made to locate the new secretariat
there as well. Although this made perfect sense to the host country, it
did present some logistical difficulties to all the international members of the Commission, who would have to meet in Hobart each
year. It would be difficult to pick another spot on the globe that is as
equally inconvenient to all the nonresident members of the Commission as Hobart, so maybe there was some sense of natural justice at
play. Thus every year for the last thirty-six years the Scientific Committee and the Commission have met for two weeks in my adopted
home town, Hobart.

The venue for the meeting has changed as the Commission accretes
new members and its requirements for space increase. What used to
be a cozy club for a few dozen Antarctic specialists has expanded
into a full-scale environmental/diplomatic roadshow involving over a
hundred scientists and even more lawyers, diplomats, fisheries managers, and directors of Antarctic institutes from twenty-five members, eleven acceding countries, and numerous observers from other
international organizations, interested nonmember nations, fishing
industry associations, and conservation groups. But what does this
annual three-ring-circus achieve?

The first meeting of the Commission was held from the May 25 to
June 11, 1982. These dates are important because the only war to have
been fought (partially) in the Southern Ocean, the ten-week Falklands War, ended on June 14, 1982. While hostilities were under way
in the South Atlantic, the major protagonists and their allies were
meeting in Australia, ostensibly to discuss krill. In the report of this
first meeting, which was largely concerned with administrative matters, there is no mention of the Falklands War, but then there was also
no progress made on krill fisheries management either. In fact, there
was no mention of krill in the reports of the first two meetings of the
Commission. It was not until 1989 that the issue of managing the krill

fishery was raised on the floor of the Commission. I still remember the apparent surprise expressed by the delegation of the Soviet Union (the largest krill fishing nation at the time) at the appearance of the topic of krill conservation on the agenda of the 1990 Commission meeting. What had happened to the krill convention?

Kick-starting an international Commission is no easy feat. There was a need to organize how the meetings would operate, how information would flow in to the decision-making body, and how essential work could occur between annual meetings. CCAMLR had also inherited several fisheries that were in a parlous state after only a few years of operation. The Commission sensibly spent its early years establishing procedures that would allow efficient decision making, developing ways to access "the best scientific information available."

From the start, the Scientific Committee was tasked with providing the much-needed scientific information on the species being fished. It became apparent that meeting once a year, in the week before the Commission meeting, was not an efficient way of compiling scientific information into a form that was understandable to the Commission. So the Scientific Committee established a series of working groups that would meet during the year between meetings to conduct calculations and amass information that the Scientific Committee could use to inform the Commission's decisions.

The first working group, established in 1984, was to provide information on the status of fish stocks. Finally, in 1989, a working group that focused on krill was formed, although it mysteriously became subsumed into a working group that focused on ecosystem monitoring and management shortly thereafter. These working groups have evolved over time and have developed offshoots and subgroups, but they are still the powerhouses of the Commission's scientific work, providing the scientific advice on which conservation measures are based. The pattern of information flow became established. The Commission (advised by scientists) indicated its priorities, the scientists went away and did their homework, reporting back first to the working groups and then to the Scientific Committee, which in turn made recommendations to the Commission, which makes all the decisions. These decisions usually take the form of Conservation Measures that bind member states to specific courses of action—for example, an annual catch limit on a species in a part of the Convention area. These Conservation Measures are agreed to by consensus. Many have perceived this requirement for agreement as a stumbling block to achieving the goals of the Convention, but this necessary agreement only relates to "matters of substance." My favorite clause in the whole Convention addresses this issue: "The question of whether a matter is one of substance shall be treated as a matter of substance."

I have now sat through three decades of rather turgid Commission meetings and there are times when I feel jaded by the experience. But, despite the seemingly endless discussions that often seem to be going nowhere, the Commission delivers the goods annually. I remain astonished that each year, on the final Friday of the annual meeting, after two weeks of arcane, intense, and often acrimonious discussions, a host of legally binding decisions are made and agreed to, in four languages, by consensus. There are few international environmental organizations that can boast this record of achievement, so a degree of tedium can probably be excused.

The requirement for consensus decision making was one of the outcomes of the Convention negotiations. Nations that were wary of the experience in the International Whaling Commission were keen to avoid the political maneuverings associated with votes that require some sort of majority. In all my years in CCAMLR I do not think I have ever seen any issue put to a vote within the Commission. Measures are discussed on the floor, but more importantly, in the margins of the meeting, with all interested parties. Generally, only Conservation Measures that are informally agreed to by all parties are then put to the Commission for final approval. This system favors negotiation over confrontation and the long game over short-term gains. If a proposal is not likely to be agreed to by all members, it is reworked in the period between meetings, discussions are held between protagonists, and a new proposal that has a higher chance of success is presented at the next year's meeting. Many initiatives take years of discussion before some sort of compromise agreement is reached, and the annual reports of the Commission can include long paragraphs of regret by Members and their allies whose initiatives were rebutted again.

~

The language used in CCAMLR is very distinctive and ritualized, and it took me several years to feel comfortable that I understood what was going on at the meetings and in the reports, and even now I sometimes struggle. This is not just the subject matter, or the way the meetings are organized, it also involves the arcane art of disagreeing strongly with a colleague while seeming to wholeheartedly support their argument. Things at CCAMLR happen slowly—but sometimes the impossible can happen quickly.

The Commission was initially tied up in frantic efforts to introduce legislation to protect fish stocks against heavy resistance from fishing nations. The effort was concentrated on obtaining protection for the species that were most at risk from continued fishing—and krill was not viewed as being in this category. The general viewpoint in CCAMLR in the 1980s was that the krill stock was large, and the reported catches were so far below the levels that had been bandied about in the 1970s that there was no immediate cause for concern. Besides, the fishing nations gave their word that their operations were not about to expand, so there was no need for any regulations.

Most of us working in CCAMLR in the late 1980s learned our first Russian word fairly quickly—*Nyet!* This initial period of sparring over fish stocks gave CCAMLR members experience in hard and lengthy negotiation. Late-night sessions in a small annex, which became known as the Wombat Room, became part of CCAMLR tradition—hence the wombat statuettes awarded for long service. I would sit through these sessions amazed that those negotiating could remain sane and even-tempered right through until four o'clock in the morning, which seemed to be the magic hour when compromise would be reached. These hardy souls would next be found at their delegation table five hours later, lucidly presenting the results of the night's discussions to the Commission. This ritual seemed an inevitable by-product of the consensus decision-making process. The early meetings of the Commission seemed like a hospital emergency department conducting triage, with the fish stocks being viewed as the most in peril and so getting the most attention. CCAMLR's ecosystem management mandate largely fell by the wayside in the rush to legislate against the further decimation of the fish.

Despite krill being central to the design of the Convention they really failed to attract the attention of the Commission until the late 1980s. In the background, the Scientific Committee proceeded with a plan of scientific work that aimed to provide the basis for measures that could be used to regulate the krill fishery. The work on krill ran in parallel with other developments that were expanding the ecosystem approach—something the krill fishery would need in the future. A program for monitoring populations of seals and penguins that were dependent on krill was established in 1989. The resultant data could be used to see if the krill fishery was having an adverse ecological effect.

This standardized data series is comparable between sites and across time. The program exists to this day and has generated valuable information on how some vertebrate species cope with the ever-changing Antarctic environment. But it has never unequivocally detected any ecological changes associated with krill fishing and has never been used in managing the krill fishery, something its instigators had hoped would happen.

In the late 1980s, it became apparent from the results of the BIO-MASS scientific survey that the astronomical levels of krill abundance that had been talked about in the 1960s and 1970s were unlikely to be realistic. If the population size of krill was smaller than we thought, then the sustainable level of harvesting would also have to be much lower. Some of the BIOMASS-era findings on the productivity of the krill stock were also entering the scientific mainstream, and these too would affect the developing approaches to krill management.

But management of the krill fishery required information from the fishery itself. When the Convention was signed, almost nothing was known about how the krill fishery operated, where it fished, and how much it caught. In the 1980s, information provided by fishing nations improved scientists' understanding of the fishery, which led to more informed discussion about management. The fishing nations steadfastly reassured the Commission that they had no intentions to expand their levels of catch, and they argued against any regulation of the fishery. Paradoxically, their argument was based on the potential economic impact of such regulations on states that might be contemplating an expansion of their fishery.

The fishing nations also pointed to the relatively small catches of krill taken when compared to the very large stock of krill, but they, like everybody else, were unable to provide reliable estimates of the stock size of krill. Finally, and possibly most significantly, they pointed to uncertainties over the ecological effects of the fishery as a rationale for postponing management action.

Such lack of action in the face of uncertainty ran entirely counter to the spirit and the wording of the Convention. In response, the Commission and its Scientific Committee began to take a more active interest in krill. In 1991, the Commission agreed that the practice of taking management action only when it was needed was not a viable long-term strategy for the krill fishery. This set the scene for the development of precautionary management of the fishery—putting in place regulations before there was an obvious need. Despite the signing of the Convention in 1981, the krill fishery in 1990 remained unregulated. An unregulated krill fishery meant that members of CCAMLR could catch as much krill as they wanted from any area of the Convention area at any time of year. This state of affairs could not continue indefinitely.

Once the Commission and the Scientific Committee turned their attention to krill, things began to happen. Between 1991 and 2009 the krill fishery moved from being an unregulated activity to a fishery that had checks and balances ensuring, for the most part, that it would operate in the manner envisioned in the Convention. The priority was to establish some limits on the tonnage of krill that could be caught in the various regions of the Convention area. The main areas of interest were those that had already been targeted by the fishery.

Establishing catch limits that were suitably precautionary to meet the ambitions of the Convention was not an easy task. A new formulation had to be arrived at that ensured the krill stock was not likely to be threatened by overfishing and that also considered the requirements of all the animals that depended on krill for food. This new approach required information on the biological condition of the krill stock: the natural rate of mortality, the annual rate of influx of young into the krill population, and the growth rate of krill. All these biological features were subjects of considerable research in the late 1980s and early 1990s. Obviously, there were many unknowns in this process, but at every step along the way these uncertainties were incorporated into the calculations. The single most crucial piece of information, however, was an estimate of the population size of krill in each of the areas to be managed.

The area under CCAMLR's jurisdiction has three broad regions: Area 48 (the South Atlantic), Area 58 (the South Indian Ocean), and Area 88 (the South Pacific Ocean). These vast areas were subdivided into smaller subareas, particularly around island groups. To set krill catch limits CCAMLR needed biomass estimations for each of these subdivisions. Surveying such huge areas was a major undertaking. Fortuitously, the results of the BIOMASS krill surveys in the early 1980s were, at last, becoming available, and these were used to set the first krill catch limit in 1991.

At the 1991 meeting of the Commission, under the highly tentative agenda item entitled "Consideration of Possible Limits on Krill Catches," the Commission broke new ground by setting the first precautionary catch limit on the krill fishery. It adopted a Conservation Measure limiting the catch of Antarctic krill in the South Atlantic to 1.5 million metric tons in any fishing season. Perhaps more importantly, the Conservation Measure also established a catch level of 620,000 metric tons a year, which cannot be exceeded until overall catch limit has been split up into smaller limits applied at a much smaller scale. This clause, designed to protect krill-feeding predators, was slipped in at the last minute during the negotiations and is now possibly the single most important element in the management of the krill fishery. Although the catch limit in this Conservation Measure has been changed several times using new data, or new scientific procedures, the 620,000 metric ton "trigger level" has remained in place unchanged.

In 1992 the Commission ruled that conservation measures in force with no time limit are understood to be in force until revoked by the Commission. This means that changing the 620,000 metric ton trigger level, which is the effective cap on the fishery, would require consensus, which would be very hard to achieve. This major achievement was the culmination of years of scientific endeavor and negotiations within CCAMLR, but sadly it was greeted with little fanfare either within CCAMLR or in the outside world.

The ball was rolling, and in 1992 the Commission agreed to another conservation measure, limiting the annual krill catch in the Southwest Indian Ocean to 390,000 metric tons, again using the BIOMASS data. No sooner had these limits been agreed upon, than the deficiencies of the BIOMASS data were being pointed out. There were rumblings about the need for a new krill survey in the South Atlantic specifically designed to provide CCAMLR with a biomass estimate that could be used to set a new catch limit. For some people the existing catch limit was set too high and for some it was too low, but everyone understood the field of krill surveying had moved on considerably since 1980. Any new survey would provide a much more accurate, precise, and up-to-date result, and probably in a much shorter time frame.

During these discussions, it was also pointed out that there was one krill fishing ground that was still being fished but remained unsurveyed—the Southeast Indian Ocean sector. This comment was aimed at the Australian delegation in the room. The Southeast Indian Ocean sector lies off the Antarctic territory claimed by Australia, and two of Australia's research stations are along this coastline. If this body of water was to be surveyed, then there was no doubt that CCAMLR expected Australia to take the lead. So began a long journey.

Surveying a large area of the southern Indian Ocean with the aim of providing an estimate of the population size of krill is a daunting task. This region is about the same area as that of the American states of Texas and Oklahoma put together. The best way to estimate krill population size is to use scientific echosounders (see chapter three), which provide a picture of how much krill there is below the survey ship. The technology and methodology for conducting these surveys had undergone considerable development during the 1980s and 1990s.

To use this information systematically, a survey must be designed to give the best chance of obtaining a snapshot of the krill population in the area. These surveys involve the ship sailing along a number of imaginary parallel lines laid out perpendicular to the coastline. The combined results of all these transects provides a map of krill distribution, and it also gives us an average density and an overall biomass that can then be used to calculate how much krill can safely be caught. Although a huge, multinational survey of the southwest Atlantic was being planned, we in Australia elected (with some considerable international encouragement) to conduct the necessary krill survey in the Southeast Indian Ocean sector.

Australia had commissioned a new research icebreaker, the *Aurora Australis*, in 1990, and we were keen to put it through its paces by tackling the ambitious task of surveying the romantically named CCAMLR Division 58.4.1. We assembled an international team of forty-two scientists with an aim both to survey krill and to provide all the information necessary to understand how this vast ecosystem worked.

Our voyage almost never left port. I arrived on the bridge of the *Aurora Australis* on the morning of January 6, 1996, the date we were due to sail, only to be informed by the ship's captain that the gearbox was broken, and it would take weeks to fix it. My heart sank, and the team stood down waiting for news about the arrival of spare parts from Europe and the repair. I spent the next few weeks in frantic negotiations trying to keep our precious voyage alive so that the four years of planning and preparation would not be wasted. Finally, some three weeks later, the parts arrived, the gearbox was repaired, and the reassembled team set sail on our epic voyage.

There is a long-standing tradition that all research voyages have a nickname and an acronym. When we reassembled on the ship we discovered that the computer programmers had secretly entered a new acronym into all the ship's databases as the ship sat idly in port. Thenceforth the voyage was named BROKE, in honor of the faulty gearbox. We spent most of the voyage trying to define this acronym, and after a fierce competition onboard we arrived at the final formulation: Baseline Research on Oceanography, Krill, and the Environment, which almost summed up what we were doing.

Once the gearbox was functioning, the voyage went off nearly perfectly. It was a long trip—seventy-two days at sea. We collected detailed datasets on everything from viruses to whales and described the oceanography along a coast that had not been surveyed before. Above all, we collected the acoustic data that would allow us to estimate the krill biomass in CCAMLR Division 58.4.1.

Conducting these big research surveys is an exercise that involves both considerable excitement and extended tedium, and not a modicum of hardship too. It took over a week to reach the start of the first transect line, just north of the ice edge at longitude 80° east, on January 29, 1996—my oldest daughter's ninth birthday. Once on survey, the ship steamed slowly up and down the eleven transect lines with all the underway instrumentation collecting data. We also stopped at predetermined intervals to lower instrumentation and nets into the water. The southern limit of the lines was always the edge of the pack ice—many of the instruments do not work well in ice—and the northern limit was at latitude 62° south, or when our sampling program told us we had moved out of krill territory. Our routine was set, sailing away from the ice for three days, turning east until we arrived at the next transect line, then heading south until we encountered the ice edge.

The part of the trip we most looked forward to was the day spent sailing across the face of the pack ice, with all its wildlife and stunning icescapes, before we turned left and began the next 130-nautical-mile northward transect. We repeated this ritual nine times over the course of the summer until we reached the edge of the CCAMLR Division on the March 29, 1996.

When we started the survey, it was high summer, and from the bridge you could watch the evening sun slowly angle toward the horizon, then skim the silhouetted icebergs before slowly rising again. By the end of the survey we were experiencing long evenings and dark nights as autumn set in. We had encountered thick ice, howling winds, and stormy waters, but mostly conditions were sufficiently benign to allow us to continue working. As we turned for home, Antarctica put on a farewell show for us as humpback whales surfaced among the sunset-tinged bergs, and as darkness engulfed us, an aurora lit the night sky. We had completed all the eleven transects and returned to port in Australia, weary after two months of relentless sampling, our computers groaning with the mass of accumulated data.

We were mindful of the decade-long complexities involved in analyzing and presenting the BIOMASS data, and our aim was to showcase how, with more modern computers and advanced data collection systems, we could go from data collection to precautionary catch limit in less than a year. This was particularly important because planning was beginning for the proposed international multiship survey of the South Atlantic, which would produce a new regional biomass estimate for the most heavily fished area.

So began a frantic three months as we collated and analyzed the data and calculated the ultimate biomass figure. Our deadline was the CCAMLR Working Group meeting in Norway in July 1996, only four months after we had disembarked from the *Aurora Australis*. The final calculations were completed only minutes before we boarded the plane to present the data to our skeptical colleagues, who did not imagine that it was possible to complete the survey and its analysis in such a short period. However, our estimate of the krill biomass for Division 58.4.1 was accepted by the Working Group and was passed on up the decision-making chain.

Decision making in CCAMLR is hierarchical; the Scientific Committee used our biomass estimate to calculate the precautionary catch limit. This limit was then recommended to the Commission, which would, we hoped, establish a Conservation Measure limiting the fishery. Finally, in October 1996, after two weeks of discussion, the Commission report dryly noted that "an Australian hydroacoustic survey in Division 58.4.1 provided a biomass estimate of 6.67 million metric tons. This was the first acoustic survey of a CCAMLR statistical division designed to produce an estimate of $B_o$ [biomass]." They then, rather gratuitously, added, "In future it would be desirable to repeat the survey so that some assessment of the variability of krill abundance in this Division could be made."

For those of us who had spent years planning and months at sea to produce this outcome, the thought of a repeat survey was disheartening. But the Commission adopted the Conservation Measure establishing a precautionary catch limit for krill of 775,000 metric tons in the Southeast Indian Ocean sector—our goal had been achieved. By the end of this process, I calculated that I had spent more time in meetings discussing the planning, conduct, and results of our survey than I had out at sea surveying.

Despite the disinterest of the Commission, the scientific work on krill continued. The main game was the massive effort to resurvey the Southwest Atlantic. So in the late 1990s considerable effort was put into planning, and finally executing, the CCAMLR 2000 survey. This was not quite as ambitious as the BIOMASS surveys and involved only four ships from four nations, Japan, Russia, the United Kingdom, and the United States, and it focused on the two million square kilometers (772,000 square miles) of Area 48 where the fishery was now concentrated. The survey was conducted in January and February 2000. The biomass estimate was presented to the July 2000 Working Group meeting in the beautiful Sicilian town of Taormina, a considerable achievement given that the survey had involved four nations and as many ships.

The estimate was 44.29 million metric tons, considerably higher than the estimate from the results of the 1981 BIOMASS program (15 million metric tons). Such a large difference between the results of both surveys was not a surprise—the extra 30 million metric tons of krill were attributed to improvements in the estimation technique over the intervening twenty years—we were getting better at detecting krill. This biomass estimate allowed a new precautionary catch limit of 4 million metric tons to be calculated for the Southwest Atlantic, and this was subsequently adopted by the Commission.

This Conservation Measure has been through several more iterations since then (the catch limit currently stands at 5.61 million metric tons), but the two essential features remain: a precautionary catch limit and the 620,000 metric ton trigger level. Both of these limits have had considerable influence on both the krill management strategy and the economic development of the fishery.

CCAMLR now had up-to-date catch limits in the South Atlantic and the Southeast Indian Ocean sectors. For the Southwest Indian Ocean sector, there was little hope of revising the catch limit based on the BIOMASS data, so in 2000 the Scientific Committee recommended resurveying this area, and once again Australia was designated the task.

After the BROKE survey in 1996, and the subsequent analysis and writing up of the scientific results, which took until 2000, I had made a definitive statement to anyone who would listen—"Never again." But in January 2006, I found myself once again sailing south on a mission to resurvey the Southwest Indian Ocean sector for CCAMLR. The voyage was unimaginatively named BROKE-West, but nothing broke this time, and the voyage was remarkable largely because of its lack of incident.

In seventy days at sea we did not lose a single day to bad weather, and all the equipment and scientific research operated flawlessly. We returned to port with another computer loaded with data and another subsequent period of intensive research, analysis, and politics. Finally, in 2007, the Commission agreed to a new precautionary catch limit for the Southwest Indian Ocean sector of 2.6 million metric tons derived from a krill biomass of 28.75 million metric tons, and the limit was geographically split into two, based on the detailed oceanographic data we had collected. As a bonus, the Conservation Measure also applied 452,000 metric ton "trigger levels" in a similar fashion to that applied in the Southwest Atlantic. Once again, our labors had yielded the required result.

The net result of over a decade of international survey work was that all the major regions of the Southern Ocean that had been fished for krill were now protected by precautionary catch limits. Some areas that were outside the regulated areas could still be fished for krill, but in 2009 a new Conservation Measure was negotiated that limits krill fisheries in unsurveyed areas to 15,000 metric tons a year and imposes a host of regulatory and compliance measures on any such fisheries. In 2009, the Commission also split the trigger level in the Southwest Atlantic geographically, to prevent all 620,000 metric tons being taken from a single small area. It was this limit for the fishery off the Antarctic Peninsula that was first reached in 2010 when the fishery was closed for the rest of the season. These two developments were almost the final pieces in the krill management jigsaw puzzle.

At the start of this process in 1991 there had been no limitation on krill fisheries anywhere in the area that the Commission managed. From 2009 onward, any krill fishery operating in Antarctic waters was subject to strict limitations. There are now nine specific conservation measures in place regulating the krill fishery, including precautionary catch limits, the level of scientific observation, data reporting, and several more general measures that also affect krill fishing operations. Several of these measures incorporate elements of the ecosystem-based management approach that was a foundation stone of the Convention.

But this remarkable achievement does not mean that the Convention has reached its objectives. A major problem remains: how to ensure that the krill fishing fleet does not take all its quota out of a small area, thus endangering the dependent wildlife. The statistical areas for which the limits and trigger levels are set are huge, and the limits apply to the entire area. The 2009 splitting of the South Atlantic trigger level into subareas is a temporary measure that does help to spread the catch out, but a more lasting approach is needed.

The wording of the Conservation Measure limiting krill catches in the South Atlantic indicates that, until the precautionary catch limit has been divided into smaller areas, the krill fishery is limited to catching 620,000 metric tons of krill each year—the trigger level. This trigger level has effectively become the real (de facto) catch limit. Reaching agreement on dividing the catch limit into smaller areas would set the scene for removing the current 620,000 metric ton cap on the fishery. It remains to be seen whether there is any appetite on the part of scientists and politicians to conduct the research that might allow krill catches to rise above this level.

Despite the obvious achievements in developing a system of management for the krill fishery, within the Commission the management of the krill fishery has never had prominence. Why has the Commission drifted away from its krill roots? The answer is not simple and has multiple causes. Initially, the fish stocks were badly damaged and needed immediate attention, but this early focus on fish has continued to this day. The processes and procedures set up in the early years and the continued engagement of scientists whose focus is on fish ensures a fishy focus, despite these fisheries being very small on a global scale.

The economics of the fisheries are viewed as being unequal. Krill is usually caught in bulk and is of low value, but fish are high-value products, and consequently there have been commercial interests in many delegations who had a vested interest in discussing fish. The high value of the fish also made them vulnerable to poaching, which occupied the Commission's attention for many years. The legal and diplomatic activity around addressing illegal fishing is the stuff of dreams to international environmental bureaucrats and lawyers. Thus the Commission has been only too willing to spend days discussing the intricacies of catch documentation schemes, port state inspections, and illegal vessel registers, to the exclusion of the complexities of the ecosystem approach to management.

In contrast, krill fishing was dull; there were no real controversial aspects to the fishery; no one appeared to want to fish for krill, legally or illegally; and besides, krill remain an enigma to most Commissioners. The fact that the Convention had been designed to manage the krill fishery has been largely forgotten.

The unfortunate collateral damage that focusing on fish has had for CCAMLR is that it has lost much of what made it unique as a Convention—the ecosystem approach. There have been attempts to salvage some of the Convention's ecosystem credentials, most recently in the focus on marine protected areas (MPAs). The focus on MPAs is a worldwide phenomenon and is seen by many as the panacea for the deterioration of many of the oceans' ecosystems.

Unfortunately, the Commission's approach has been largely based on politics rather than on an assessment of what needed to be protected from what and where. The first large MPA to be established in CCAMLR-managed waters is situated in the Ross Sea, a region where krill fishing has rarely occurred. Protecting the ecosystems of the Southern Ocean from the threats of overharvesting is a laudable goal, but without a doubt, the largest potential threat comes from an expanded krill fishery. Yet the Commission has turned away from addressing this threat. This is not to say that the krill fishery is un-managed; but rather, there has not been a planned or systematic approach to ecosystem management as was initially envisaged.

Although CCAMLR has formal responsibility for managing the krill fishery, an interesting recent development is the emergence of third parties that are seeking to accelerate the pace of progress toward comprehensive krill management. The Marine Stewardship Council (MSC) is an independent body that assesses the sustainability of fisheries, and in recent years two krill fishing companies have received certification. This certification often comes with qualifications, and these can stipulate more stringent controls than CCAMLR has instituted. Having an MSC certification is seen as important, particularly to companies that are selling highly visible products, such as krill oil, so complying with MSC requirements is given a high priority by the industry.

In the last round of certifications, the MSC pointed out that the management system for krill is based on a biomass survey that was conducted in 2000 and that, in their opinion, future certification will depend on having a more up-to-date estimate of the abundance of krill on the fishing grounds. Plans are now afoot by several CCAMLR member nations, and by the krill fishery, to conduct a new survey in 2019. Significantly, this initiative is being developed outside CCAMLR, and the krill fishing industry is a major supporter of, and a likely participant in, this project.

The coordinated approach by the krill fishing industry was made possible by the formation in 2011 of the Association of Responsible Krill Fishing Industries (ARK), which aims to assist the krill fishing industry to work with CCAMLR to ensure the sustainable management of the fishery. Because ARK only has observer status to CCAMLR it has to operate outside the formal structures, but this makes it possible to react much more rapidly than the Commission. Through ARK, the krill fishing fleet has been able to take unilateral action in cases where the ponderous deliberations of the Commission cannot reach agreement.

In 2016, ARK vessels volunteered to stay away from some Gentoo penguin colonies that were suffering poor breeding success due to extensive ice cover. CCAMLR had deliberated such action but, despite lengthy discussions, was unable to decide on the correct wording for a resolution that would have enacted a voluntary exclusion zone around the penguin colonies. ARK's decisive action was a welcome surprise to the Commission and to environmental nongovernmental organizations and may reflect a sense of impatience in the krill fishing industry that desires some progress on krill management and seeks some long-term certainty.

These examples indicate that, in the future, these external bodies may have an increasing influence on the management of the krill fishery—without the need for the Commission to engage in long-winded discussions prior to reaching consensus. I suspect that future management of the krill fishery will involve not only the best scientific advice from CCAMLR's Scientific Committee but also direct action by a krill fishing industry that is keen to be viewed as a model of sustainability. Such a hybrid approach may mean that significant progress toward the Convention's goal of ecosystem-based management of the krill fishery may occur, irrespective of whether the Commission sees it as a priority.

So has the Convention succeeded or has it failed? From my perspective, it is too early to tell. The innovative aspects of the Convention will only be properly tested if the krill fishery grows to the size where it might begin to affect ecosystem processes. Paradoxically, if the Commission is truly successful, then it will be able to manage the krill fishery in a way that skillfully avoids affecting ecosystem processes.

Judged against the goals it set itself from the outset in Article II of the Convention, the results so far are mixed. For example, population sizes of some harvested fish have fallen below optimal levels. The "ecological relationships between harvested, dependent, and related populations of Antarctic marine living resources" have been disturbed and have not recovered. Depleted populations of some vertebrates have recovered—but these species fall under the purview of other international conventions dealing with whales and seals. Some species of seabirds are now protected from mortality associated with fishing thanks to the Convention, but their populations were largely threatened as a result of fisheries sanctioned by the Commission in the first place.

Finally, has the Convention resulted in the "prevention of changes or minimization of the risk of changes in the marine ecosystem which are not potentially reversible over two or three decades," as stated in Article II of the Convention? Some management measures have allowed overharvested species the breathing space to recover, but even now, some two or three decades later, many have not shown the anticipated recovery.

However, on the positive side, we are undoubtedly in a far better place now than we would be had the Convention not been implemented. Some conservation groups have argued, often for the wrong reason, that the Convention has failed and should be scrapped or replaced. CCAMLR is, however, the one environmental conservation treaty that covers most of the living organisms in the Southern Ocean and on the Antarctic continent, so its very existence is extremely valuable. Negotiating a replacement would be a herculean task, but developing a new Convention that was as full of good concepts, novel approaches, and lofty aspirations would be nigh on impossible.

In the late 1970s an article was published in the *Times of London* describing the negotiations that would result in the CCAMLR Convention; the Southern Ocean was described as "an ideal natural laboratory in which to test a political application of ecological principles on a scale that has not been tried before." This experiment into science (and politics) continues to this day.

*Chapter Eight*

# Krill Futures

I fear the worst too, but only because fearing the best is an
absolute waste of time!
—Will the Krill in *Happy Feet 2*

I have rarely seen a more miserable sight. A tiny, shivering Gentoo penguin chick stood dripping in the rain on the guano-stained rocks of Peterman Island. This drenched native of the frozen continent was as unused to Antarctic rain as I was. I would survive the experience, but it was unlikely that this poor undersized individual would. There are winners and losers in the changing climate. Gentoos are believed to be one of the beneficiaries of the warming that has so affected the Antarctic Peninsula, and they are spreading south as the sea ice retreats, whereas their neighbors, the Adélies, withdraw to more icy conditions. Adélies are in decline in the South Atlantic, but they flourish further south, whereas Gentoo numbers are on the increase in the north. These population changes are undoubtedly climate related, though some have tried to link them to changes in the population of krill, or the influence of the krill fishery, or the increase in fur seal numbers. And therein lies the problem. There are multiple environmental changes occurring simultaneously so how can we disentangle cause and effect, and how do we dispassionately assess causation?

I may well have the world's largest collection of newspaper headlines related to krill. Admittedly, it is not a huge collection, but they all follow a similar theme. "Lack of *Krill* a Big Worry for hungry Penguins;" "Warming Oceans May Threaten *Krill*, a Cornerstone of the Antarctic Ecosystem;" "Life in Antarctica Relies on Shrinking Supply of *Krill*;" "Climate Change Could Put Tiny *Krill* at Big Risk;" "Acid Oceans Could Kill *Krill*;" and "Ecologists Fear Huge Rise in *Krill* Catch." These headlines are usually followed by the familiar but misguided phrase; "Krill are tiny shrimplike crustaceans that form the base of the Antarctic food-web," before the bulk of the text selectively quotes some recent research that spells doom for krill, but more importantly, for the animals that depend on krill.

We all know that bad news grabs the headlines. We should not be surprised, then, when most of the news that features krill stems from reports prophesizing doom and gloom. Scientists want to, or are often forced to, channel their research into topics that address issues of environmental concern, such as climate change, pollution, or overfishing. This channeling is a result of funding agencies developing strategic plans that, rightly, address the highest-priority topics, and these are usually topics related to human-induced change in its many forms. So we end up with scientists being funded to address issues associated with ecological threats. The reports published from this research inevitably have to focus on the bad news (or at least feature the bad news in the report titles). The media then pick up on the press releases with catchy headlines from the institutes that have sponsored the research.

Most research that is conducted on any ecosystem produces findings that can be spun to illustrate either a good news story or a bad news story—depending on the whims of a subeditor. I learned this a long time ago. After we had surveyed the East Antarctic ecosystem in 1996, we submitted the data to the Commission for the Conservation of Antarctic Marine Living Resources (CCAMLR, hereafter referred to as the Commission), so they could set a catch limit on the fishery in the area. Following the adoption of a catch limit on krill for

A waterlogged Gentoo penguin chick shivering in the unexpected rain on the Antarctic Peninsula. (Photo by author)

this area, the headline in a national newspaper was "Going in for the Krill: Australia Is to Allow Fishing for Antarctic Krill, the World's Last Big Untapped Food Resource." In fact, what had happened was that our research had actually produced the results that were used to *limit* fishing for krill in an area where the fishery, at the time, was unregulated. The text of the article revealed the good-news truth, but the headline proclaimed a bad-news story, or at least a message that would be interpreted that way by a large number of readers. This is the world we live in, but the take-home point is that there is more to any story than the sound bites or the newspaper headlines.

There is a great need for scientists to get the entire story out to the general public so people can develop their own informed opinions—or seek other sources to verify the story they have been told. Journalists need to move beyond sensational headlines and learn to explore the subtleties of science. Scientific research is not simple. When it involves complex ecosystems in regions of the world where access is difficult, the story is always incomplete, and the results have to be interpreted carefully. This puts a huge responsibility on scientists to be evenhanded in their approach to controversial subjects—and that is asking a lot—we are, after all, just ordinary people. But scientists do not have to work alone.

Many scientists are now collaborating with artists, writers, musicians, and filmmakers to broaden the audience for complex scientific stories. Moving images of living krill capture the imagination of viewers, and artists can convey the fluid and rhythmic movements of krill as well as their aesthetic qualities. More complex artwork can interpret concepts like environmental change and can position the subjects of our scientific study in a more accessible medium. Although art, to many scientists, is viewed as an interesting adjunct to more serious studies, there is a growing trend for scientists to work with artists to achieve wider recognition for the wonders of the natural world. Many artists are luring scientists into their exotic world, and both groups find their collaborations deeply rewarding. Being hard-nosed about it, scientific publications reach audiences in the hundreds, at best. Artists' exhibitions can draw tens of thousands of people from all walks of life, including rich and influential people—precisely the

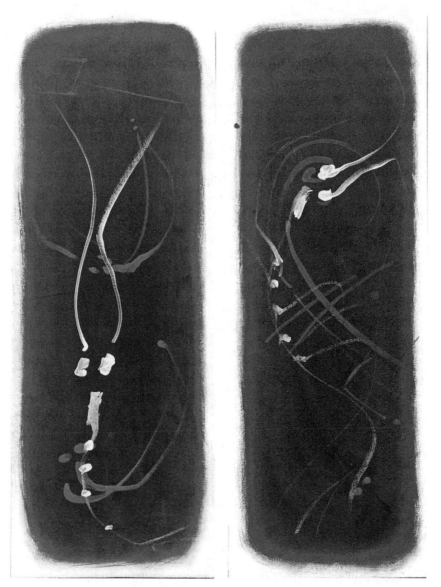

Some rare examples of krill art drawn from observations of living krill by Lisa Roberts. (Photos by author)

size and type of audience ecologists would like to reach with their message. Krill research is becoming more inclusive, and artistic interpretations mean that their curious life can be admired and respected by a much larger and broader audience.

Now, after decades in the wilderness, krill are gaining some small amount of public recognition. They are seen more frequently in documentaries and even in animated feature films—their popular cultural pinnacle was probably reached in the animated film *Happy Feet 2* where, for once, they played a starring role. Greater recognition leads to more questions being asked, and hopefully this leads to a better understanding of krill and to more research being carried out to improve that understanding. This understanding should translate into more informed discussion around krill and the conservation of their ecosystem into the future. But what does the future hold for krill?

To understate the case, the future for krill is uncertain. As I write this, nuclear posturing in the Northern Hemisphere threatens to make all life on the planet uncertain. Assuming we survive this current crisis, we will have to address the wide range of environmental changes that are affecting all life on the planet. There is a tendency to view environmental change as a threatening process, but for adaptable animals, like krill, it may end up being challenging rather than life threatening. Krill are among the world's most abundant animals. Over millennia they have thrived, rather than being threatened by every change in their challenging physical and biological world, but how they have done this is yet unknown.

The history of krill in the Antarctic is opaque. In yet another confounding quirk of krill, we have been unable to find any fossil record for any species of krill that would assist us in learning how they have reacted to historical changes. We know that Antarctic krill probably emerged about twenty million years ago, but how the population responded to the tumultuous changes that have occurred since then is a mystery. We do know that krill are remarkably successful so they must have considerable powers of adaptation, and this gives me some hope for their future.

The main problem, however, is that so many changes are now happening rapidly and simultaneously. Animals can adapt to slow change over evolutionary periods, and animals with short life spans can adapt the fastest. Unfortunately, because krill are long-lived, their ability to adapt to rapid change might be limited. But they do have one weapon for avoiding the worst element of a changing ocean—behavior. By reacting to an unfavorable environment krill can alter their distribution and stay away from areas that are threatening. This avoidance behavior is probably only limited, but coupled with slow physiological adaptation to change it may allow krill to survive even in areas where their doom has been predicted.

As I have emphasized, krill behavior is one of the great unknowns. Future studies examining the fate of the krill population when faced by a warming, more acidic ocean with less sea ice will need to incorporate a realistic assessment of their behavioral abilities. There are currently few data that could be used to do this. Those who are brave enough to try and predict the effect of a warming ocean on the krill population have a fiendishly complicated task.

It has usually been assumed that krill are merely being acted upon by their physical environment. We have already seen that not taking krill behavior into account can lead to errors in our predictions. We know that krill migrate both vertically and horizontally, and there is some evidence that when the surface waters are too warm, they can spend more time in the cooler deep water. So krill are unlikely to be passive players in the game of global warming.

This is not to assert that there are not immense environmental challenges to the survival of krill, and by extension, the Southern Ocean ecosystem. We know that changes to critical aspects of the Southern Ocean are occurring right now and have in the recent past. It is uncertain how krill can cope with these changes, and it is also not immediately evident what we can do at the local level to ensure the conservation of krill.

Large-scale global changes are already affecting the Antarctic region. The atmosphere over the Southern Ocean is warming in step with global trends, or in some cases ahead of them. The Antarctic Peninsula is the fastest-warming area on the planet, and the increased air temperature over this region has resulted in a host of physical and biological changes, including increasingly damp penguins. Terrestrial animals are having to adapt to a new environment; some will cope, others will migrate to regions where the conditions are more favorable.

For krill, cocooned in a cold ocean, the warming of the atmosphere will still have significant effects. The warmer air, together with a warmer ocean, contributes to faster glacial melt rates, and this will locally reduce the salinity of the surface layer of the ocean. Krill can cope with some reduced salinity, but they may avoid such areas, which could affect their distribution. A further complication of melting glaciers is that they deposit vast quantities of finely ground rock into the water. This "glacial flour" is so fine it remains suspended in the water and can clog the filtering apparatus of krill. This glacially induced constipation has been associated with mass krill strandings on the island beaches of the Antarctic Peninsula. These consequences of a warmer Antarctica may be highly localized, but they might well drive the krill population deeper or further offshore in the Antarctic Peninsula region. Should the krill move away from the colonies of seals and penguins that depend on them, the ecosystem will suffer.

On a larger scale, temperature changes in the atmosphere affect circulation patterns. Antarctica is surrounded by belts of strong winds, which are driven by differences in the density of air between the continent and the ocean. These winds mix the ocean and move the sea ice around. Stronger than average winds are becoming more prevalent and have been shown to compact sea ice against the western Antarctic Peninsula, increasing rafting and average ice thickness. Like so many elements of environmental change, this works in favor of some animal species and is a detriment to others.

Ocean warming is the sleeping giant of climate change. Much of the excess heat retained by the planet because of the carbon dioxide greenhouse effect has ended up in the vast bulk of the ocean. Because of its size the ocean responds slowly, and the changes are more difficult to observe.

There are, however, good records showing a steady warming of the deep ocean right around Antarctica. Warming has also been observed in the critical surface layer. Near the Antarctic Peninsula, the summer temperature in surface waters has increased by 1°C (1.8°F) over the last fifty years. At the northern limits of krill distribution, around

South Georgia, the temperature of the surface layer has increased 0.9°C (1.6°F) over the last eighty years. Projections suggest that further widespread warming of up to 1°C (1.8°F) will occur by the late twenty-first century. Such a significant warming of the ocean—remember krill only tolerate seawater temperatures between − 1.8°C (− 35.2°F) and 5°C (41°F)—is predicted to reduce the area of the suitable habitat for krill by twenty percent. This range shrinkage may be offset because the warming ocean may stimulate more algal growth, which would mean more food for krill. Even though the temperature may no longer be optimal for krill growth, they still might be able to maintain their population growth rates, but in a smaller area, if the mix of food available remains favorable.

The warming of the ocean is an indirect effect of the carbon dioxide that has entered the atmosphere because of the burning of fossil fuels. The excess carbon dioxide in the atmosphere eventually finds its way into the surface of the ocean, and once there it makes the water more acidic. This ocean acidification is occurring right now. The colder water of the Southern Ocean takes up carbon dioxide more rapidly than the warmer waters to the north. So the habitat of krill is acidifying more quickly.

An acidic ocean will particularly affect the ability of animals to make shells out of calcium carbonate. A range of crustaceans, mollusks, and smaller organisms will struggle in a more acidic ocean. Krill have only lightly calcified shells so this change may not affect them greatly from this perspective. But an acidic environment affects a whole range of physiological processes. Laboratory studies have shown that krill embryonic development stalls under extremely acidic conditions. These experimental studies have been incorporated into predictive ocean models, and it looks like krill reproduction may not be possible in important krill habitats, such as the Weddell Sea, within one hundred years. This is the bad news. The good news is that adult krill appear to be resilient to increased acidity, so the net effect of ocean acidification will be a trade-off between the different life history stages. Once again, behavioral adaptations will probably play a major role in determining the net effect of acidification on Antarctica's pivotal species.

It is not difficult to see the link between increases in the temperature of the ocean and the atmosphere and changes in sea ice extent, concentration, and thickness. Thanks to the advent of satellite imagery, we know that the Antarctic sea ice distribution and concentration are quite different now from what we were observing in the 1980s. The seasonality has also changed—ice is advancing later in the autumn and retreating earlier in the spring. It is difficult to put these changes in a historical context because our records do not go back very far; satellite imagery has only been available since the 1970s.

The satellite record does show us that the overall amount of sea ice formed around the Antarctic continent in winter, in terms of area of ocean covered by ice, has not changed dramatically since the 1970s, but on a regional scale there are quite distinct differences. Around East Antarctica and into the Ross Sea there may have been an increase in the area covered by sea ice during winter. In the South Atlantic, however, there has been a significant decrease in the winter sea ice area, to the extent that some locations are now perennially ice-free. This is a worrying development because the South Atlantic is home to the largest populations of krill, and there have been links demonstrated between krill productivity and sea ice extent.

But, as always, there are winners and losers from the sea ice changes. Krill adults might grow better in the presence of more extensive open-water algal blooms, but their larvae need a supply of ice algae, and the refuge that sea ice provides, to overwinter successfully, and this habitat may be in decline. Decreases in sea ice could spell bad news for seals with fewer sites for breeding. In contrast, penguins may benefit because they will have a shorter walk from their continental colonies to their open-water feeding grounds. Both penguins and seals would have better year-round access to open water where they can feed, which could put more pressure on the krill populations.

Increased pressure could also come from the krill fishing fleet, which can now operate year-round. However, this spreads out the fishing effort and means that the fleet is not always competing with krill predators when they most need it—to feed their offspring in summer. But the longer season may well interest new fishing companies, which may increase the fishing pressure on the whole system.

Whereas it is intuitive to grasp the potential effect of changes in the sea ice and the ocean on the krill population it is less apparent how atmospheric changes can directly impact the Antarctic ecosystem, which is largely marine. However, in the 1990s the discovery of the "ozone hole" in the stratosphere above the Antarctic continent initiated a flurry of research. The lack of ozone was exposing Earth's surface to increased levels of ultraviolet light. UV radiation can be very damaging to organisms. It is responsible for sunburn in humans, and it can disrupt critical metabolic processes in both plants and animals. UV-B is the most damaging form of ultraviolet radiation but, critically, it is rapidly absorbed in the top twenty meters (twenty-two yards) of the seawater. This offers a measure of protection to marine organisms. But the air-breathing animals of the region can spend much of the year on the ice, in colonies ashore, or on the water's surface, so they may have been exposed to increased radiation for the last thirty years. If there had been an intense ecological response to the opening of the ozone hole it would have happened in the 1970s, and the populations we can observe now are those that have adapted to the new conditions.

Initially, we thought that krill, because of their migration patterns, would be protected against the increases in dangerous radiation. Once again, krill proved us wrong; the makeup of their DNA makes them genetically predisposed to damage by UV-B. But krill have several strategies they could have employed to cope with increased UV levels. They take up specific molecules from their algal food that serve as sunscreens; they can change their behavior—spending more time in deeper water or altering their vertical migration pattern would shield them from the harmful rays. Like the other Antarctic animals, krill may have evolved to deal with the new UV levels—less resilient strains would have died off while more robust strains would have flourished. We will probably never know what effect increased UV-B levels have had on krill, or the rest of the Antarctic ecosystem. Hopefully the measures taken to halt the production of chemicals that damage the ozone layer will allow the levels of UV to return to those of previous centuries.

Some of the most profound changes that will occur in the Southern

Ocean over the next few decades may come not from physical changes in the ocean and atmosphere but from the recovery of formerly over-exploited species. The removal of the great whales over such a short period has been described as a "huge uncontrolled ecological experiment," Currently, the great whales are on their way back, with some populations of species, such as humpbacks, recovering as fast as is biologically possible—up to an eleven percent increase in numbers per year. This huge-scale ecological perturbation is happening rapidly, yet there are precious few resources going into understanding the imminent effects of this recovery. If all species of krill-eating whales start to recover as rapidly as the humpbacks then the krill-based ecosystem is in for rapid change, and this has implications for all other vertebrate species in the Southern Ocean, for the ability of the ocean to absorb carbon dioxide and for the region's fisheries.

The sustainability of the krill fishery and its impacts on the rest of the ecosystem is a matter of concern to a great number of people. At its current level, there is no doubt that the fishery is sustainable—the catch is less than 0.1 percent of the estimated global krill biomass. To most people, though, the question is more about whether the current and future catches can harm the rest of the ecosystem. That potential can exist if the fishery concentrates in too small an area; thus spreading out the krill fishery is a key goal within the Commission on the Conservation of Antarctic Marine Living Resources, which has done more than any other international agreement to ensure that all its fisheries are sustainable.

There will always be those who would wish there were no krill fishery at all. This is one potential future, but it is unlikely to be realized, so the alternative has to be a well-regulated fishery that is responsive to observed changes in the ecosystem. Recent developments suggest that the krill fishing industry has recognized the economic advantages of operating in a fishery that is acknowledged as being sustainable, so fishing operators may become the leaders in pressing for comprehensive management of the fishery. These developments are promising given the Commission's tardiness in enacting such a management approach. Consequently, I am cautiously optimistic about the effect of future harvesting of krill, assuming that the practice is

well-regulated through an international treaty with sustainable principles that all fishing nations and companies agree to and adhere to.
I also assume that fishing nations and companies will operate on a
sound economic model, whereby fishing will happen only if the costs
of the operation are met by the revenues obtained from the sale of
krill-based products. Unfortunately, the history of the krill fishery
tells us that this is not always the case.

Fisheries for many valuable species all around the world have been
plagued by pirate fisheries—officially termed illegal, unreported,
and unregulated (IUU) fisheries. These fisheries operate outside the
agreed regulatory framework and sell their products on the open
market, often under different names to hide their criminal origin.
This scourge has not yet affected the krill fishery, and most experts
assume that it won't because there is no open market for krill; fishing
operators catch to supply an existing demand. Krill are also a low-
value product, but gearing-up to go fishing, particularly to produce
premium krill-based items, is very high cost.

IUU fishing for Antarctic species, such as toothfish, can be carried
out by relatively small, cheap, and available long-line vessels. White-
fleshed fish (including toothfish) are also extremely valuable, so the
rewards for the risks of pirate fishing are very high. A good catch of
toothfish could pay off the cost of buying a long-liner in a single trip.
Contrast this with the cost of accessing the sort of huge supertrawler
necessary to go krill fishing, and having to spend months on the fishing grounds to make any sort of economic sense. Besides, the current
catch of krill is so far below the regulated catch limit that there is no
benefit to operating outside the management system. If you want to
go fishing for krill, it makes sense to do it legally, and this also makes
it easier to obtain the necessary finances.

All of that said, things change rapidly, so complacency is not
an option. The best approach to guaranteeing that the krill fishery
remains under adequate control is to ensure that the Commission
is working effectively and that the Antarctic Treaty System under
which CCAMLR operates continues to be the agreed method for
governing the region. The key to conserving krill will come from the
world of diplomacy not from scientific research.

Management of the krill fishery is one of the few "levers" that we have to affect the future of the Southern Ocean ecosystem. We can increase or decrease catches of krill knowing that this will have some effect on the system, however small. Other changes in the system are global in nature and cannot be effected by simple management actions in one sector of the economy or region. Because of this level of control over fishing activity, there will be increasing pressure on CCAMLR over the years to come, because it is the only international body with the authority to implement measures that can affect the whole ecosystem. When presented with evidence of declines in penguin populations, could the Commission, given its conservation remit and precautionary approach, avoid taking restraining action on the krill fishery—even if it would have little ecological effect?

In this book, I have emphasized the gaps in our knowledge and lamented the primitive state of our research tools. But I do think that the next few decades will see a renaissance in research into krill and other marine organisms. We will never achieve the degree of certainty that physicists aspire to, but with careful study and the application of new technology, we will be able to make great strides in marine biology. In the 1970s, when I was investigating krill swarms in the Bay of Fundy, I longed for some form of technology that could accurately determine our location, see beneath the boat, and observe the shape of swarms from above. Today, only forty years on, I could achieve all of this with a mobile phone, an off-the-shelf echosounder, and a mail-order drone—with little cost. Given these technological advances, I imagine I could complete the studies that took me several years within only a few days. The availability of all the new "toys" is a liberating force that will move our understanding of the ocean forward at a rapid pace.

Simple acts, such as capturing images of krill on underwater video, have changed the way in which we view the vertical range of krill in the ocean and have given new insights into their behavior. And this is from just a few reported observations. James Marr, back in the 1960s, also foresaw a future where observation augmented more conventional sampling:

Perhaps, it is the bathyscape, or some such contrivance, from which we can *watch* as well as capture the animals we have so long been blindly sampling, that we can obtain a reliable picture of their natural abundance, as well as much else concerning them that at present remains to us obscure. (Marr, 1962, 264)

There have been very few attempts to take a manned submersible—the bathyscape envisioned by Marr—to the Southern Ocean, but miniature cameras, under-ice scuba diving, and remotely operated vehicles are improving our observational power.

Satellites, aircraft, drones, and autonomous underwater vehicles can now be used to map krill populations and to follow their movements, but we still have difficulty keeping track of individual krill. One of the most significant developments in studying the biology of larger animals, both on land and in the ocean, has been the development of equipment that allows researchers to track individuals in space and time. A range of miniaturized sensors have been developed that provide information on the location of an animal, how fast it is moving, what depth it is at, and what it is feeding on, as well as providing the temperature and salinity of the ocean. Such instrumentation has revolutionized our understanding of the behavior of large marine animals.

Our understanding of how krill *predators* operate in their natural environment has increased dramatically over the last two decades; why has this revolution not flowed on to krill? There are two main reasons. First, diving mammals and birds must visit the surface of the ocean to breathe, and flying birds spend most of their lives above the waves. This means that signals from instruments that have been attached to them can be relayed to satellites. Second, even though miniaturization has developed at an amazing pace, we are not yet at the stage where we can attach a sensor to an aquatic animal the size of a krill. I suspect that with the current size of most sensors we would find that a krill, loaded down with an electronic backpack and released into the ocean, would head straight to the ocean floor and lie there until some benthic predator made a meal of it. For us to benefit from the electronics revolution we will need much smaller instruments and an efficient way of sending information through water. I am sure these developments will happen soon, and it will probably transform our science the way that satellite tracking has changed the way we think of vertebrate predators.

Back in the laboratory, it is the genetic revolution that holds great promise for the future. Two decades ago there was little conception of the revolution that was about to be unleashed through developments in the field of molecular biology. This area of research is developing so fast that it has outstripped the ability of "normal" biologists (like me) to be able to keep up and understand even what tools are available. Luckily, a younger generation of scientists are now beginning to apply their skills and knowledge to many of the intriguing issues surrounding the biology of krill and the management of the krill fishery.

Molecular techniques are being applied to individual krill and to populations of krill, but, predictably, there are problems thrown up by the enigmatic creatures themselves. A significant obstacle for geneticists is the huge genome of krill—twelve times the size of our own. This means that, even in the era of rapid and cheap genome sequencing, it will take a few more years of technical development before the krill genome can be sequenced and we can begin to unravel its secrets. But molecular techniques have already been used to determine what krill eat by looking at the DNA in their guts and how the timing of the krill life cycle is regulated through an internal clock. Given the pace of the molecular revolution, it is impossible to predict what might happen next.

The key to applying all these new methodologies to answer the critical questions concerning krill is to ensure that our key questions are widely appreciated so that those with the appropriate skills are attracted to the scientific challenge of answering them. But it is not enough to merely apply new methods; we must also develop new concepts and a new language to describe the world that our technology opens for us. There is a subtle and overlooked interaction between the techniques we use to study life in the ocean and the language we use to describe its inhabitants and their interactions. Victorian terms such as *zooplankton* need to fade away as we discard Victorian tools like the plankton net. The next few decades will be an exciting time to be a krill researcher and will change forever the way we view krill and their environment—I hope I can be around to observe this revolution.

As we gain more insights and observations from using new technology our knowledge of krill and its environment will improve immeasurably. In parallel, mathematical models will be developing as computing power increases and as clever mathematics percolates its way into the field of ecology. The combination of improved observations, better biological understanding, and enhanced modeling power will mean that in the decades to come we will be in a far better position to make some meaningful predictions about the future state of the Southern Ocean ecosystem and of the animal that is at its center. In turn, these predictions will allow us to make more considered decisions about management of activities in the Antarctic region, including fishing. This does not mean that we should wait until our knowledge and our methods are perfect (they never will be) before acting. Rather it suggests that we should exercise extreme caution now, and only when we have greater certainty can we relax our guard somewhat.

The conservation of krill is critical for the future of the Southern Ocean ecosystem. Such a statement assumes that there is consensus on the qualities of the ecosystem that should be conserved. The CCAMLR Convention went some way toward defining the principles of conservation that govern harvesting activities. In a changing environment, and with limited data, it is difficult to know which form of the ecosystem should be the preferred baseline and how to ensure that human actions drive the system toward this preferred state. There is an oft-repeated myth that the Antarctic region is pristine— we know that there has been a two-hundred-year history of extreme exploitation. The current ecosystem is quite different, physically and biologically, from that encountered by the first humans who explored the Southern Ocean nearly three hundred years ago. Do we want to restore the ecosystem to its preexploitation state? If so, how would we do this? Is continued exploitation of krill and fish compatible with the restoration of the preexploitation ecosystem? These questions are more philosophical than scientific and can only be answered through well-informed political actions.

With the catalogue of environmental changes already under way I fear that the political process will be unable to respond fast enough, and we may all sit on the sidelines while the planet's ocean ecosystems change. The best that we can do is to develop our ability to monitor the change and improve our capability to predict the future state of the ecosystem and spread the message widely about the state of the environment. Better information is our best weapon in the fight to ensure the conservation of the Antarctic region.

Despite my reservations, I remain cautiously optimistic for the future of krill in the Antarctic. Krill is an extraordinary animal. Its abundance, size, nutritional content, and swarming behavior have ensured its critical ecological importance. Its longevity, physiological plasticity, and behavior have all contributed to its major role in many of the ocean's biological, chemical, and physical cycles.

Yet, despite these aspects, all of which contribute to resilience, we must be extremely cautious when scoping out potential futures for the krill population. Some of the planet's most abundant species, such as the passenger pigeon, were exterminated over very short periods of time, so mere abundance alone does not guarantee survival against a backdrop of natural and human-induced changes.

We are now at a critical point in history for krill—our actions now could spell doom for an entire ecosystem, or we could make some well-informed choices that enhance the conservation of krill and thus the entire Antarctic ecosystem. In contrast to Will and Bill, I am prepared to fear for the best.

I do hope that in the pages of this book I have managed to convince you that krill is indeed the superb krill and deserves concerted conservation efforts to ensure future welfare of its population. It is an amazing animal that is vitally important in the Antarctic ecosystem and is a species of global significance. By extension, other species of krill in all the world's oceans are similarly amazing and deserve the attention of scientists and all others who care for the health of our oceans.

Furthermore, the myriad of smaller animal species that inhabit the watery seventy percent of our planet need to be taken more seriously and be brought firmly into the public consciousness. We need to utilize the new technology that is becoming available to reinvent our thoughts about life in the ocean and to draw public attention to the wonder and complexity of marine ecosystems. A focus on large marine animals—the "charismatic megafauna"—helps to draw the public's attention to the threats that face the oceans, but we do need to develop a better public understanding of how the oceans work. This will require broadening our consciousness about the teeming hordes of life in the oceans that are currently unknown and unseen and whose song remains unsung. Perhaps krill can become a pinup species for a new category that can capture the public attention—"the mesmerizing mesofauna"—and this will assist us to focus on the less obvious but undoubtedly critical elements of the marine ecosystem. I hope so.

# *Acknowledgments*

I would like to express my eternal appreciation to my family, who have endured my many absences as I headed south in search of my chosen crustacean. In particular, my wife, Dianne, who has been with me every step of the way on my thirty-six-year voyage into the wonderful world of krill, deserves my special thanks, if not a medal.

A special round of applause goes to my small team of proofreaders—daughters Sarah and Charlotte, and Angela McGaffin—all of whom found many of my deliberate mistakes.

Erin Johnson of Island Press tortured me until I abandoned my scientific upbringing and started writing proper sentences in understandable English. I am truly grateful for her persistence and assistance.

Danielle Wood and Heather Rose wonderfully orchestrated a creative writing course at the University of Tasmania, which gave me the confidence to tackle this project. Elizabeth Leane has encouraged my writing habit and has been my guide in making the transition from pure science to a more holistic view of the universe.

Small passages of chapter two also appeared in an essay I contributed to "Antarctica: Music, Sounds and Cultural Connections," published by ANU Press in 2015, and are reproduced with the permission of the editors. The use of excerpts from *Happy Feet II* is granted courtesy of Warner Brothers Entertainment Inc. Thanks also to Marion Stoneman of the Tasmanian Writers' Centre who helped me to make some sense of the world of publishing.

I am forever indebted to Ron O'Dor, whose faith in me (and in krill), allowed me to complete my studies at Dalhousie University and who taught me so much about marine animals. Similarly, Ray Thurber, the Nova Scotian fisherman, guided me in my first stumbling steps in studying the ocean. His knowledge, humor, wisdom, and tolerance made those early years messing about on a boat in the Bay of Fundy such a pleasure.

The crew of the *Aurora Australis* and vast numbers of my scientific colleagues spent months of sea with me as I pursued my krill fantasies. To my great surprise and pleasure many of them are still among my best friends.

Finally, I am extremely grateful to the Australian Antarctic Division, which nurtured me through twenty-four years' hard labor and tolerated my undisciplined approach to science, policy, and administration. They have also generously allowed me to use some of their magnificent images in this book.

# Glossary

**A-frame.** An A-shaped structure at the rear of research and fishing vessels. Winches haul cables connected to nets through pulleys on the A-frame.

**ARK (Association of Responsible Krill Fishing Industries).** This body, composed of most of the current krill fishing companies, aims to assist the krill fishing industry to work with the Commission for the Conservation of Antarctic Marine Living Resources to ensure the sustainable management of the fishery (see the ARK website: http://www.ark-krill.org/).

**ATS (Antarctic Treaty System).** The range of agreements that have developed from the Antarctic Treaty and which provide a system of governance for Antarctica.

**Biomass.** The scientific term for the mass of animal life. It is usually calculated from the number of animals in a volume of water multiplied by their average weight.

**BIOMASS (Biological Investigations of Marine Systems and Stocks).** An international research program of the late 1970s and early 1980s that sought to provide a comprehensive overview of the ecology of the Southern Ocean. BIOMASS had a focus on krill and produced the first large-scale estimates of krill abundance.

**CCAMLR.** This acronym is used for two related entities: The Convention on the Conservation of Antarctic Marine Living Resources (the international agreement that is responsible for fishing and conservation in the Southern Ocean) and the Commission for the Conservation of Antarctic Marine Living Resources (the body that oversees the Convention). The Commission has twenty-four members and eleven acceding states (nations that have observing rights but not voting rights at Commission meetings). Details about the Convention and the Commission can be found at the Commission's website: https://www.ccamlr.org.

**CCAS (Convention on the Conservation of Antarctic Seals).** An Antarctic Treaty System body that was established to regulate commercial sealing in the Antarctic. It is currently dormant because there have been no recent applications to hunt seals in the Antarctic.

**Cod-end.** The collecting device at the rear of a net into which the catch collects. The cod-end on scientific nets is usually small (less than five liters) but fishing vessels use cod-ends that can collect up to ten metric tons of catch. Scientific nets have a detachable cod-end, and the contents of the collecting device are usually emptied into a bucket or other receptacle. Commercial fishing nets have a fixed cod-end, and there is some form of opening mechanism so that the contents of the net can be emptied into the hold of the vessel.

**Discovery Investigations.** A series of scientific cruises and shore-based investigations named after the British research ship the *RRS Discovery*. These investigations were intended to provide the scientific background to management of the Antarctic whale fishery. Thirty-seven volumes of reports were produced between 1929 and 1980. Two volumes established the foundations for all later research on Antarctic krill: *The Natural History and Geography of the Antarctic Krill* (Euphausia superba Dana) by James Marr, published in 1962, and *The Life Cycle of Antarctic Krill in Relation to Ice and Water Conditions* by Neil Mackintosh, published in 1972.

**Echosounder.** An instrument used to measure the distribution and abundance of animals in the water. Some echosounders are also used to examine variations in the seafloor topography. Echosounders work by sending out a pulse of sound from a transducer. The sound is reflected from objects in the water, and the reflected sound is then captured by the transducer. The strength of the reflected sound beam is a product of the body composition of the animals and how densely packed they are. Because sound is absorbed by water, echosounders work best when close to their targets, and they begin to lose their effectiveness when the targets are deep in the water. The effective depth of operating depends on the sound frequency put out by the echosounder. Many modern scientific echosounders use several beams of different frequencies, and the differences in returned signal can be used to discriminate between species in the water.

**Euphausiid.** The scientific name for a family of eighty-five species of shrimplike marine crustaceans, commonly known as krill, that are found throughout the oceans of the world. Most Euphausiid species are free-swimming, or pelagic, animals.

**FAO (Food and Agriculture Organization of the United Nations).** The international body that collects and analyzes data on the world's fisheries.

**Fast ice.** Sea ice that is attached to the Antarctic continent and which does not move with the currents.

**Formalin.** A solution of the chemical formaldehyde diluted in seawater. It is a commonly used chemical to preserve catches of marine animals for research.

**Genome.** The complete set of genes or the genetic material present in a cell or organism.

**Gills.** The structures of aquatic animals that exchange gas (oxygen, carbon dioxide) with the surrounding water. Gills are the aquatic equivalent of lungs.

**IGY (International Geophysical Year).** This was the first internationally coordinated approach to Antarctic physical sciences and occurred in 1957/1958. The international cooperation it fostered is credited with the push to develop the Antarctic Treaty, which now governs activities in Antarctica.

**IWC (The International Whaling Commission).** This body is responsible for the regulation of whaling in all oceans.

**Krill.** The common name used to describe all members of the group of crustaceans known as Euphausiids. The term is often used to specifically refer to the most abundant species—Antarctic krill—known scientifically as *Euphausia superba*.

**Marine Stewardship Council (MSC).** A fisheries certification body whose mission is to use their ecolabel and fishery certification program to contribute to the health of the world's oceans by recognizing and rewarding sustainable fishing practices, influencing the choices people make when buying seafood, and working with partners to transform the seafood market to a sustainable basis (website: https://www .msc.org/).

**Metazoan.** Animals whose body is made up of many differentiated cells. This term is used to distinguish complex animal life from simpler unicellular animals.

**MPA (Marine Protected Area).** MPAs are areas of the ocean that have been set aside either to allow recovery from previous exploitation or because of their particular scientific values. Activities such as fishing are either banned or highly restricted in MPAs. The world's largest MPA was established in the Ross Sea in Antarctica in 2017.

**Pack ice.** Ice formed from the freezing of seawater. Most pack ice forms and melts each year, though some is multiyear ice, which is correspondingly thicker. Pack ice drifts with the currents and extends for twenty-two million square kilometers (eight and a half million square miles) around the Antarctic in winter. In summer the ice recedes to cover only two and a half million square kilometers (965,000 square miles).

**Pelagic.** A term that refers to animals of the open ocean. These animals are free-swimming, unlike many marine species that live on or at the seafloor.

**Phytoplankton.** A term to describe the huge variety of floating unicellular plant life in the ocean. These plants, which I have also termed algae in the text, are the base of the Antarctic food chain and are responsible for producing much of the oxygen we breathe. They absorb carbon dioxide and produce oxygen through the process of photosynthesis. Because they are unicellular, phytoplankton can bloom rapidly when conditions are right—sufficient nutrients, sunshine, and a mechanism for keeping them close to the water's surface.

**Polynya.** A large expanse of ice-free water in the middle of the sea ice zone in winter. Polynyas are usually caused by strong winds in the interior of Antarctica blowing the ice offshore faster than it can refreeze.

**SCAR (Scientific Committee on Antarctic Research).** This body is part of the ATS and is designed to coordinate scientific activities in Antarctica.

**Swarms and schools.** Swarms are very densely packed animal aggregations. Schools are also composed of densely packed animals, but all the animals are facing the same way with a regular distance between all individuals in the school. Schools move through the water, but they can form when animals are stationary in a current. Swarms may merely be stationary schools, but they also form when krill are feeding.

**Thermocline.** A density boundary between the surface layer of the ocean and deeper waters. The boundary can be a result of differences in temperature and salinity. The seasonal thermocline is broken down in winter when storms mix up the surface waters.

**Zooplankton.** A term used loosely to describe most aquatic animals that are below average size. They are assumed to be passively drifting and unable to affect their location by swimming. The term encompasses animals that feed on plants, bacteria, other animals, and detritus so has no real basis in ecology. It also covers all known taxonomic groups of aquatic animals.

# Further Reading

There is very little of a nontechnical nature written specifically about krill. This recent multiauthored book on all aspects of the biology and ecology of Antarctic krill, including the krill fishery and its management, can be consulted for a summary of the most up-to-date science: *Biology and Ecology of Antarctic Krill*, edited by Volker Siegel (Springer, 2016).

I highly recommend taking a journey through James Marr's magnificent *The Natural History and Geography of the Antarctic Krill* (Euphausia superba Dana) (Discovery Reports 32, pages 33–464, Cambridge University Press, 1962) if you can find a copy. The beautiful quote from Ommaney in chapter one is found on page 156 of this volume. The whaler's 1892 account of whale and krill sightings in chapter two appears on page 151. Mackintosh's 1934 quote in chapter three is taken from page 258.

An accessible introduction to all aspects of Antarctic science can be found in *Antarctica—Global Science from a Frozen Continent*, edited by David Walton (Cambridge University Press, 2013).

I explored the potential for krill and large animals to affect the fertility of the ocean in an article, "Givers of Life," in *New Scientist*, July 9, 2011, pages 36–39.

There is some amazing film footage of krill in two BBC series, *Life in the Freezer* and the *Frozen Planet*. A recent documentary film, *License to Krill*, produced by David Singleton, also has some good footage and some interesting discussions about the krill fishery and the topic of krill population changes.

# About the Author

Stephen Nicol is a scientist who has spent his entire professional life working with krill in the Antarctic, as well as in Canada and South Africa. Nicol was born in Ireland and had an eclectic education in England, the United States, Scotland, and Canada. He has published extensively on many aspects of krill biology, on the management of the krill fishery, on the Southern Ocean ecosystem, as well as on more esoteric aspects of human interactions with krill. He worked for the Australian Government's Antarctic Division for twenty-four years as a research scientist and as a program leader. During that time, he also served on Australia's delegation to the international commission that manages Antarctica's fisheries. He was awarded the Australian Antarctic medal for his services to Antarctic research in 2011. He led four major voyages to Antarctica and has participated in four others. In 2011 he retired and enrolled in a master's degree program in creative writing, which resulted in several published short stories, photographic essays, and travel articles—and this book. He supervises graduate students and gives lectures at the Institute of Marine and Antarctic Studies at the University of Tasmania, where he holds the title of adjunct professor. He provides advice on krill to bodies such as the Association of Responsible Krill Harvesting Companies (ARK), to various conservation organizations, and to anyone who will listen.

# *Index*